Intelligent Sensor Design
Using the Microchip dsPIC®

Intelligent Sensor Design
Using the Microchip dsPIC®

by Creed Huddleston

AMSTERDAM • BOSTON • HEIDELBERG • LONDON
NEW YORK • OXFORD • PARIS • SAN DIEGO
SAN FRANCISCO • SINGAPORE • SYDNEY • TOKYO

Newnes is an imprint of Elsevier

Newnes

Newnes is an imprint of Elsevier
30 Corporate Drive, Suite 400, Burlington, MA 01803, USA
Linacre House, Jordan Hill, Oxford OX2 8DP, UK

 Recognizing the importance of preserving what has been written,
Elsevier prints its books on acid-free paper whenever possible.

Library of Congress Cataloging-in-Publication Data

Application submitted

British Library Cataloguing-in-Publication Data
A catalogue record for this book is available from the British Library.

ISBN-13: 978-0-7506-7755-4
ISBN-10: 0-7506-7755-4

For information on all Newnes publications
visit our website at www.books.elsevier.com

06 07 08 09 10 10 9 8 7 6 5 4 3 2 1

Printed in the United States of America

Working together to grow
libraries in developing countries

www.elsevier.com | www.bookaid.org | www.sabre.org

ELSEVIER   BOOK AID
            International   Sabre Foundation

This book is lovingly dedicated to my incredible wife Lisa
and my three wonderful children, Kate, Beth, and Dan.
We are truly blessed to be a family, a fact I reflect upon often.

The book is also dedicated to my sister Sarah, whose tremendous
laugh I'll always remember and who probably would have been
stunned (but hopefully pleased) to see her name here.

Contents

Introduction

Just as no book covers all subjects, neither is a single book appropriate for all readers. As much as I wish it were otherwise, *Intelligent Sensor Design Using the Microchip dsPIC* is no exception. The book primarily targets three groups of readers:

1. the experienced embedded system designer who is trying to get up to speed on the Microchip dsPIC DSC relatively quickly and doesn't have time to wade through a huge amount of detailed documentation in order to get a system up and running,
2. the engineering student who is new to embedded systems and needs extra guidance on how all the development pieces fit together, and
3. anyone with a programming or engineering background who is interested in learning more about the fascinating field of intelligent sensing.

To get the most out of the material presented here, the reader should be familiar with the basics of the C programming language. While none of the examples use esoteric C statement constructions, the focus in the text is on applications, specifically on the design of sensing systems, not on the programming language per se. If the reader does not have a background in C, there are a number of excellent books that teach the language well, and the reader is advised to consult those first.

Neither is the book a primer on digital signal processing (DSP). Although it offers a quick (one chapter) overview of the basic concepts of DSP, this is not a rigorous academic treatment of the subject. That's not meant in any way to be dismissive of such texts; they are extremely valuable, and over the years I've bought a lot of them! Nonetheless, in the book we use DSP as a tool and assume that the reader either already has some facility with the tool or can acquire it through other means. Since the goal of the material is to develop an intuitive understanding of the DSP principles that we might want to use, a deeply detailed exegesis of the subject would simply waste time. There are others far more qualified than the author to write such a book, and the accompanying CD offers some suggested reading.

Finally, the book is not a hardware design manual; although we discuss some circuitry in the examples, the focus is really more on establishing a basic signal-conditioning platform and then using the software capabilities of the dsPIC DSC to extract useful information from the systems we're monitoring. As with programming and DSP, there are a great many excellent treatises on the intricacies of analog and digital circuit design that are written by masters in the field.

Having said what the book is *not*, we're now ready to look at what the book *is*. The goal of the material is to enable readers to quickly assemble a software and hardware development platform and to assimilate the knowledge that allows them to easily experiment with various approaches to the design of intelligent sensors. This new class of sensing device is rapidly displacing traditional sensors by offering improved measurement capabilities, the ability to easily interface to a wide variety of monitoring and control systems, and other features that simply aren't available through standard sensors. To that end, the book develops three complete intelligent sensor applications in the area of temperature measurement, load cell monitoring, and flow sensing. Well-commented source code for all three applications is included in the accompanying CD, as well as links to a variety of valuable Internet-based resources.

One issue that's important is that, in the book, I use a variety of Microchip hardware and software development components, and at first blush it may appear that the book is a thinly disguised marketing effort by Microchip. Nothing could be further from the truth. Microchip has offered no financial support for this project, and while my company, Omnisys Corporation, is an authorized Microchip consultant, we are also authorized consultants for a number of other semiconductor companies, including Cypress, Freescale, and Lattice.

The primary reason for using Microchip development components is that they are readily available, inexpensive (often free), and there are user forums available to get help when the inevitable questions or problems arise. For instance, the Microchip C compiler (student edition) and its MPLAB Integrated Development Environment (IDE) are available for download for free from the company website, as is the Filter Lab 2.0 software for designing analog anti-aliasing filters (more on this later). In addition, Microchip offers dsPICWorks (a DSP analysis tool) for free and its dsPIC Filter Design Software for a very reasonable price. Most developers I know don't want to have to spend huge amounts of money just to get to the point where they can test out a few concepts. Microchip does a great job of providing low-cost and free tools. These are tools that I use in my daily development work, so I'm familiar with them.

I hope that you enjoy this book as much as I have enjoyed writing it. My experience with embedded systems, particularly hard real-time control and communication systems stretches back over 20 years, and during that time I've been fortunate to be involved with a number of very intelligent, insightful individuals working on some very difficult problems. If I've done my job well, you'll take away some of those insights as well.

Acknowledgments

The genesis of this book was an article I wrote that appeared in the January 2003 issue of *Sensors Magazine*. Titled *Digital Signal Processing Turns Thermocouples Into Superstars*, the article caught the eye of Carol Lewis, then an acquisitions editor for Butterworth-Heinemann/Newnes, a division of Elsevier Press. She graciously asked whether I would be interested in writing a book-length treatment of dsPIC-based intelligent sensors. Although Carol subsequently retired before I completed the manuscript, in one of those pleasant twists of fate, she was available to copyedit the final draft. For Carol's initial belief in my writing skills and for her subsequent efforts in shaping the final draft, I am extremely grateful.

I was also very fortunate in the form of my new editor, Tiffany Gasbarrini, who assumed the role of mentor, friend, shield, and occasionally goad (but only when absolutely necessary). Without her skill in all four roles, this book would never have seen the light of day, as both she and my wife can attest. Tiffany was ably assisted by Michele Cronin, who joined the team about six months before completion of the manuscript. Michele's attention to detail and her pleasant personality contributed both to the technical quality of the book and to the enjoyment I got out of writing it.

Over the years, I've been privileged to work with a number of outstanding engineers who helped shape my understanding of the practice of engineering design, particularly for hard real-time systems. Four in particular stand out, and you the reader will benefit directly because of what I've learned from them. Dale DuVall, John Bateson, and Dale Redford, whom I first met at Scantek Corporation in the late 1980s taught me the value of thoroughly understanding the physical processes underpinning any electronic system. While all three were the best pure engineers that I've worked with, each had his own area of expertise and provided different insights into the design process.

Dale DuVall was a physicist by training, and his understanding of how things worked down to the most detailed level was nothing short of phenomenal. Through his uncanny ability to relate similar effects in different areas of study, I learned the importance of keeping one's eyes open and not ignoring anomalies. In John, I met absolutely the best analog designer I've ever known, and one of the most intelligent and most enjoyable people as well. John's attention to detail, to doing things the right way, is still the gold standard in my eyes. Last but certainly not least is Dale Redford, a brilliant digital designer who will not settle for anything less than the best when it comes to circuit or system design. All three of these guys are special, and I remember my time at Scantek fondly.

The fourth engineer also happens to be a close personal friend, my business partner at Omnisys Corporation for the past eleven years, and hockey linemate: Fred Frantz. Together we have been fortunate to design control and communication systems that are used to manufacture products that run the gamut from neonatal heart monitors, to car bumpers, to products that steer high-end yachts electronically. Through all of it, I've learned a lot from Fred and have very much enjoyed the many hours we've spent together.

Of course, all that design work at Omnisys didn't get done by just two folks, and I'd like to thank Parry Admire, John Brashier, and Mary Frantz (yes, that's Fred's wife and an excellent engineer to boot), and Steve Gibson in particular, for both the friendship and great engineering skills they bring to the table.

Finally, though, my greatest acknowledgements of thanks have to go to those who mean the most to me: my family. Only as a parent have I come to realize just how much time and love Mother and Dad gave me growing up, and how much my sisters Susan and Sarah put up with having me as a brother. When Susan married James Belote, I finally got the brother I never had growing up; he (and now their children) always brings great light to any family gathering. When I married Lisa, my family expanded with the addition of a great mother-in-law and a great father-in-law, Johnie and Gerhard Schulz, who are truly a second mom and dad to me as well. Marrying Lisa, however, was far and away the most intelligent decision I've *ever* made, and I'm grateful every day that she's my partner in life. As for our three children, Katie, Beth, and Dan, they are truly the lights of our lives, and proof (if any was needed) of just how fortunate I am.

About the Author

With over twenty years of experience designing real-time embedded systems, Creed Huddleston is Vice President of Omnisys Corporation, a new product development company based in Raleigh, NC that specializes in the creation of hard real-time instrumentation, control, and communication systems. One of the company's founders, he is responsible for new product design and for the development of Omnisys' authorized consultant relationships with companies such as Microchip Technologies, Freescale Semiconductor, Lattice Semiconductor, and TrollTech.

In addition to his duties with Omnisys, Creed also serves on the Advisory Board of Quickfilter Technologies Inc., a Texas-based company producing mixed-signal integrated circuits that provide high-speed analog signal conditioning and digital signal processing in a single package.

A graduate of Rice University in Houston, TX with a BSEE degree, Creed performed extensive graduate work in digital signal processing at the University of Texas at Arlington before heading east to Raleigh, NC to start Omnisys Corporation. To her great credit and his great fortune, Creed and his wife Lisa have been married for 23 years and have three wonderful children: Katie, Beth, and Dan.

Creed's technical interests focus on the development and application of intelligent sensing systems, particularly in the areas of precision instrumentation, sensor networking, and high-reliability communications in low-power, long-life embedded systems. He can be reached at *creedh@intelligentsensordesign.com*.

What's on the CD-ROM?

The software on the included CD consists of an on-disk website with links to valuable resources on the Internet and the source code and project files for the three applications developed in the book. To view the website, either use Windows Explorer™ (the file management program, not to be confused with Internet Explorer, which is a web browser) to find the file index.htm in the root directory of the CD and double-click on the file. That should start your web browser and load the first page of the site. Alternatively, you can enter:

```
D:\index.htm
```

in the address bar of your web browser and press the "Go" button in the browser to load the first page. Note that this assumes that your CD drive is drive D; if this is not the case, simply substitute the appropriate letter in the path.

Appendix A provides more information about the software included on the CD-ROM.

What Are Intelligent Sensors, and Why Should I Care about Them?

We took what was a luxury and turned it into a neccesity.
—*Henry Ford*

In today's instant-access world, people want and expect to be able to get information when they want it, in the form they need, and at a price they can afford (preferably free). As Peter Drucker, the greatest management mind of the past 100 years, points out, unlike physical products, information doesn't operate under the scarcity theory of economics (in which an item becomes more valuable the less there is of it); on the contrary, information becomes more useful (and valuable) the more there is of it and the more broadly it is disseminated.[1] Individuals and organizations that understand this concept have begun to unlock the tremendous value that has lain fallow in commercial, academic, and nonprofit enterprises throughout the globe by digitizing their mountains of raw data, analyzing it to create meaningful information, and then sending that information via standardized communication links to others within and outside their organizations to accomplish meaningful work.

It would be difficult to overstate the effects that this new paradigm has wrought in society economically, intellectually, and in everyday life. People now speak about working in *Internet time*, a frame of reference in which both space and time are greatly compressed. Information from anywhere on the globe can be distributed to virtually anywhere else quickly and reliably, and Bangalore is now as close to New York City as Boston. In this new world, people work differently than before; individuals or groups can easily team with others from around the corner or around the globe to produce new ideas, new products, or new services, creating fabulous new wealth for some and destroying ways of life for others. Truly, the new paradigm, which author Thomas Friedman refers to as the "flattening" of the globe, represents a tectonic shift in the way people view their world and interact within it.

Interestingly, a nearly identical though largely unnoticed sea change is occurring in the rather mundane world of sensors. For the uninitiated, *sensors* (or *sensing ele-*

ments as they're sometimes called) are devices that allow a user to measure the value of some physical condition of interest using the inherent physical properties of the sensor. That's quite a mouthful for a pretty simple concept, namely monitoring the behavior of one (relatively) easy-to-observe parameter to deduce the value of another difficult-to-observe parameter.

An example of a very familiar nonelectronic temperature sensor is the mercury bulb thermometer, in which a column of mercury contracts or expands in response to the temperature of the material to which it's exposed. In this case, the physical condition that we're measuring is the temperature of the material in which the thermometer is inserted, and the inherent physical property of the sensor that we use for measurement is the height of the mercury in the thermometer.

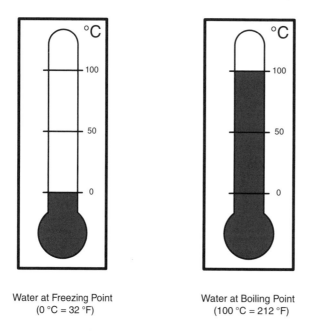

Water at Freezing Point
(0 °C = 32 °F)

Water at Boiling Point
(100 °C = 212 °F)

Figure 1.1. Two Mercury Bulb Thermometers Showing the Temperature of a Material (Ice and Boiling Water) under Two Different Conditions

So what kinds of parameters can we measure with sensors? The answer is quite a lot, actually, with the limiting factor generally being our imaginations. Probably the most widely measured parameter is temperature, but other applications include pressure, acceleration, humidity, position, pH, and literally thousands more. What makes sensors so useful, though, is not just their ability to accurately measure a wide range of parameters but that the sensors can perform those measurements under environmental conditions in which human involvement is simply impossible. Whether it's measuring the temperature of molten steel at the center of a blast furnace or monitoring the ocean current thousands of feet below the surface, sensors provide the accurate information that allows us to monitor and control all sorts of important processes.

At first glance, it might seem that sensors fall in the same category as a comfortable sweatshirt, nice to have but not particularly exciting. In this case, such a first impression would be dead wrong. To put things in perspective, in 2005 there were an estimated 6.4 billion people living on the planet.[2] Coincidentally, the market for industrial sensors in the United States alone in 2005 was estimated to be $6.4 billion,[3] and $40 billion worldwide. There are far more sensors in the world than humans, they're called upon to do tasks that range from the mundane to the cutting edges of science, and people are willing to pay for the value that sensors bring to the table. That's a powerful and profitable confluence of need, technical challenge, and economic opportunity, and into the fray has stepped a new class of devices that is bringing disruptive change to the sensing world: *intelligent sensors.*

Just what are these intelligent sensors? Conceptually, they're a new class of electronic sensing device that's literally revolutionizing the way we gather data from the world around us, how we extract useful information from that data and, finally, how we use our newfound information to perform all sorts of operations faster, more accurately, safer, and less expensively than ever before. Even better, we can leverage the power of individual intelligent sensors by communicating their information to other intelligent sensors or to other systems, allowing us to accomplish tasks that weren't possible before and creating incredible advancements in a wide variety of applications. Sound familiar?

1.1 Conventional Sensors Aren't Perfect

Before we delve into a discussion of intelligent sensors, we first need to examine regular sensors a bit more closely so that we have a solid foundation upon which to develop our understanding of intelligent sensors. For all that they do well, most sensors have a few shortcomings, both technically and economically. To be effective, a sensor usually must be *calibrated*—that is, its output must be made to match some predetermined standard so that its reported values correctly reflect the parameter being measured. In the case of a bulb thermometer, the gradations next to the mercury column must be positioned so that they accurately correspond to the level of the mercury for a given temperature. If the sensor's not calibrated, the information that it reports won't be accurate, which can be a big problem for the systems that use the reported information. Now, not all situations require the same level of accuracy. For instance, if the thermostat in your house is off by a degree or two, it doesn't really make much difference; you'll simply adjust the temperature up or down to suit your comfort. In a chemical reaction, however, that same difference of a degree or two might literally mean the difference between a valuable compound, a useless batch of goop, or an explosion! We'll discuss the issue of calibration in greater depth later, but for now the key concept to understand is that the ability to calibrate a sensor accurately is a good, often necessary, feature. It's also important to understand that, as important as it is to calibrate a sensor, often it's extremely difficult if not impossible to get to a sensor in order to calibrate it manually once it's been deployed in the field.

The second concern one has when dealing with sensors is that their properties usually change over time, a phenomenon known as *drift*. For instance, suppose we're measuring a DC current in a particular part of a circuit by monitoring the voltage across a resistor in that circuit (Figure 1.2).

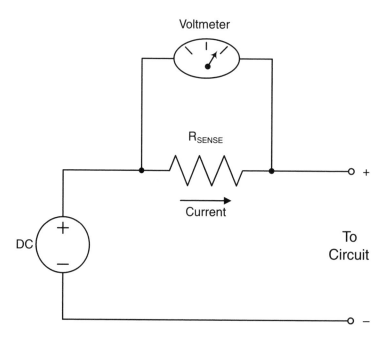

Figure 1.2. Example of a Resistive Sensing Element Used to Measure Current

In this case, the sensor is the resistor and the physical property that we're measuring is the voltage across it, which, as we know from Ohm's Law,[4] will vary directly with the amount of current flowing through the resistor. As the resistor ages, its chemical properties will change, thus altering its resistance. If, for example, we measured a voltage of 2.7V across the resistor for a current of 100 mA when the system was new, we might measure a voltage of 2.76V across it for the same current five years later. While 0.06V may not seem like much, depending upon the application it may be significant. As with the issue of calibration, some situations require much stricter drift tolerances than others; the point is that sensor properties will change with time unless we compensate for the drift in some fashion, and these changes are usually undesirable.

By the way, are you wondering why in the previous example I referred to the resistor as the sensing element and not the voltmeter used to measure the voltage? The distinction is a bit pedantic but important. In the example, I said that we were monitoring the current in the circuit by measuring the voltage across the resistor. That made the resistor the primary sensor and the voltage across it the property that changes in response to a change in the parameter of interest. The voltmeter is a secondary sensing device that we use to measure the primary parameter. As one might guess, the voltmeter itself has its own issues with calibration and drift as well. The reason that the distinction between the primary and secondary sensors is important is that it's critical to know precisely what you're measuring. Without a clear understanding of the parameter(s) of interest, it's possible to create a system that doesn't really measure what you want or that introduces excessive problems with accuracy. We'll devote more attention to that particular aspect later.

A third problem is that not only do sensors themselves change with time, but so, too, does the environment in which they operate. An excellent example of that would be the electronic ignition for an internal combustion engine. Immediately after a tune-up, all the belts are tight, the spark plugs are new, the fuel injectors are clean, and the air filter is pristine. From that moment on, things go downhill; the belts loosen, deposits build up on the spark plugs and fuel injectors, and the air filter becomes clogged with ever-increasing amounts of dirt and dust. Unless the electronic ignition can measure how things are changing and make adjustments, the settings and timing sequence that it uses to fire the spark plugs will become progressively mismatched for the engine conditions, resulting in poorer performance and reduced fuel efficiency. That might not strike you as particularly important if you're zipping around town and have a gas station on most corners, but you probably wouldn't be quite so sanguine if you were flying across the ocean and had to make it without refueling! The ability to compensate for often extreme changes in the operating environment makes a huge difference in a sensor's value to a particular application.

Yet a fourth problem is that most sensors require some sort of specialized hardware called *signal-conditioning circuitry* in order to be of use in monitoring or control applications. The signal-conditioning circuitry is what transforms the physical sensor property that we're monitoring (often an analog electrical voltage that varies in some systematic way with the parameter being measured) into a measurement that can be used by the rest of the system. Depending upon the application, the signal conditioning may be as simple as a basic amplifier that boosts the sensor signal to a usable level or it may entail complex circuitry that cleans up the sensor signal and compensates for environmental conditions, too. Frequently, the conditioning circuitry itself has to be tuned for the specific sensor being used, and for analog signals that often means physically adjusting a potentiometer or other such trimming device. In addition, the configuration of the signal-conditioning circuitry tends to be unique to both the specific type of sensor and to the application itself, which means that different types of sensors or different applications frequently need customized circuitry.

Finally, standard sensors usually need to be physically close to the control and monitoring systems that receive their measurements. In general, the farther a sensor is from the system using its measurements, the less useful the measurements are. This is due primarily to the fact that sensor signals that are run long distances are susceptible to electronic noise, thus degrading the quality of the readings at the receiving end. In many cases, sensors are connected to the monitoring and control systems using specialized (and expensive) cabling; the longer this cabling is, the more costly the installation, which is never popular with end users. A related problem is that sharing sensor outputs among multiple systems becomes very difficult, particularly if those systems are physically separated. This inability to share outputs may not seem important, but it severely limits the ability to scale systems to large installations, resulting in much higher costs to install and support multiple redundant sensors.

What we really need to do is to develop some technique by which we can solve or at least greatly alleviate these problems of calibration, drift, and signal conditioning. If we could find some way to share the sensor outputs easily, we'd be able to solve the issue of scaling, too. Let's turn now to how that's being accomplished, and examine the effects the new approach has on the sensor world.

> While sensors come in a variety of flavors (electronic, mechanical, chemical, optical, etc.), we'll focus in this book on electronic sensor devices, for the simple but powerful reason that we can interface their outputs to a computing element (usually a microprocessor) easily. It's the computing element that allows us to add intelligence to the sensor and, as we'll see, that's a very valuable addition.

1.2 First Things First—Digitizing the Sensor Signal

When engineers design a system that employs sensors, they mathematically model the response of the sensor to the physical parameter being sensed, they mathematically model the desired response of the signal-conditioning circuitry to the sensor output, and they then implement those mathematical models in electronic circuitry. All that modeling is good, but it's important to remember that the models are approximations (albeit usually fairly accurate approximations) to the real-world response of the implementation. It would be far better to keep as much of the system as possible actually in the mathematical realm; numbers, after all, don't drift with time and can be manipulated precisely and easily. In fact, the discipline of *digital signal processing* or *DSP*, in which signals are manipulated mathematically rather than with electronic circuitry, is well established and widely practiced. Standard transformations, such as filtering to remove unwanted noise or frequency mappings to identify particular signal components, are easily handled using DSP. Furthermore, using DSP principles we can perform operations that would be impossible using even the most advanced electronic circuitry.

For that very reason, today's designers also include a stage in the signal-conditioning circuitry in which the analog electrical signal is converted into a digitized numeric value. This step, called *analog-to-digital conversion*, *A/D conversion*, or *ADC,* is vitally important, because as soon as we can transform the sensor signal into a numeric value, we can manipulate it using software running on a microprocessor. Analog-to-digital converters, or ADCs as they're referred to, are usually single-chip semiconductor devices that can be made to be highly accurate and highly stable under varying environmental conditions. The required signal-conditioning circuitry can often be significantly reduced, since much of the environmental compensation circuitry can be made a part of the ADC and filtering can be performed in software. As we'll see shortly, this combination of reduced electronic hardware and the ability to operate almost exclusively in the mathematical world provides tremendous benefits from both a system-performance standpoint and from a business perspective.

1.3   Next Step—Add Some Intelligence

Once the sensor signal has been digitized, there are two primary options in how we handle those numeric values and the algorithms that manipulate them. We can either choose to implement custom digital hardware that essentially "hard-wires" our processing algorithm, or we can use a microprocessor to provide the necessary computational power. In general, custom hardware can run faster than microprocessor-driven systems, but usually at the price of increased production costs and limited flexibility. Microprocessors, while not necessarily as fast as a custom hardware solution, offer the great advantage of design flexibility and tend to be lower-priced since they can be applied to a variety of situations rather than a single application.

Once we have on-board intelligence, we're able to solve several of the problems that we noted earlier. Calibration can be automated, component drift can be virtually eliminated through the use of purely mathematical processing algorithms, and we can compensate for environmental changes by monitoring conditions on a periodic basis and making the appropriate adjustments automatically. Adding a brain makes the designer's life *much* easier.

As we'll see in Chapter 3, a relatively new class of microprocessor, known as a *digital signal controller* or *DSC*, is rapidly finding favor in products that require low cost, a high degree of integration (i.e., a great deal of functionality combined into a single semiconductor chip), and the ability to run both branch-intensive[5] and computationally intensive software efficiently. Although usually not as fast as custom digital hardware, in many cases DSCs are fast enough to implement the necessary algorithms. At the end of the day, that's all that really matters.

1.4   Finish Up with Quick and Reliable Communications

That leaves just one unresolved issue: sharing sensor values so systems that have to share sensor outputs can scale easily. Once again, the fact that the sensor data is numeric allows us to meet this requirement reliably. Just as sharing information adds to its value in the human world, so too the sharing of measurements with other components within the system or with other systems adds to the value of these measurements. To do this, we need to equip our intelligent sensor with a standardized means to communicate its information to other elements. By using standardized methods of communication, we ensure that the sensor's information can be shared as broadly, as easily, and as reliably as possible, thus maximizing the usefulness of the sensor and the information it produces.

1.5 Put It All Together, and You've Got an Intelligent Sensor

At this point, we've outlined the three characteristics that most engineers consider to be mandatory for an intelligent sensor (sometimes called a *smart sensor*):

1. a *sensing element* that measures one or more physical parameters (essentially the traditional sensor we've been discussing),
2. a *computational element* that analyzes the measurements made by the sensing element, and
3. a *communication interface* to the outside world that allows the device to exchange information with other components in a larger system.

It's the last two elements that really distinguish intelligent sensors from their more common standard sensor relatives (see Figure 1.3), because they provide the abilities to turn data directly into information, to use that information locally, and to communicate it to other elements in the system.

Essentially, intelligent sensors "flatten" the sensor world, allowing sensors to connect to other sensors nearby or around the globe and to accomplish tasks that simply weren't possible prior to their development. Just as importantly, because so much of their functionality comes from the software that controls them, companies can differentiate their products merely by changing the configuration of the software that runs in them.

This has two very important consequences for suppliers of intelligent sensors. First, it essentially moves the supplier from a hardware-based product to a software-based product. While it's certainly true that there has to be a basic hardware platform for the sensor (this is, after all, a physical device), the hardware is no longer the primary vehicle for adding (or capturing) value; the software that controls the intelligent sensor is. Because the manufacturer can add or delete features by flipping a configuration bit in software, it can alter its product mix almost instantaneously, and the specific product configuration doesn't have to be finalized until just before final test and shipment. One hardware platform can be used on multiple products targeted for different market segments at different price points; and, once new features have been developed, no additional production costs are required in order to include them in the product, so marginal profit soars.

The second consequence is that, because the intelligent sensor is connected to the outside world, the supplier now has the ability to gather information on the operation of its sensors in the field under real-world conditions and to update the software running the sensors after they leave the factory. Not only does the infor-

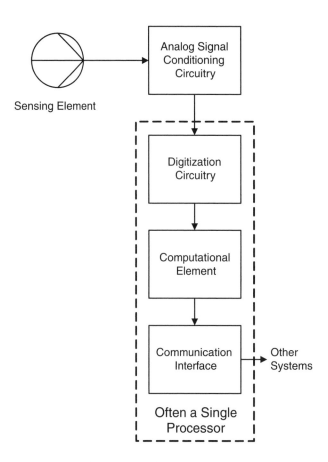

Figure 1.3. Block Diagrams of a Standard Sensor (above) and of an Intelligent Sensor (below)

mation from the field offer the sensor manufacturer unparalleled insight into the needs and concerns of its customers, but it also provides the hard data required to determine the issues that are most important to those customers (and hence are the ones that the customers are most likely to value). Armed with this information, sensor manufacturers can quickly add new features, offer certain configurations on an as-needed basis, or perform maintenance, all without having to touch the sensor itself. Services can now be delivered cost-effectively from central locations, providing yet another opportunity for the supplier to capture additional value and profits. An example of this is reported in the *Harvard Business Review*:

> Most manufacturers cannot charge more than $90 to $110 per hour for their technical support because of price and benefit pressures from local competitors. But GE Energy, because of its efficient network-enabled remote servicing, can charge $500 to $600 per hour for the same technician. Even more important, the information generated by its continual monitoring allows GE to take on additional tasks, such as managing a customer's spare parts inventory or providing the customer's and GE's service and support personnel with complete access to unified data and knowledge about the status of the equipment.[6]

1.6   Why Don't We Make Everything Intelligent?

With all of the benefits that come from turning a standard, stand-alone sensor into a connected, intelligent sensor, are there any reasons why we wouldn't want to make all sensors intelligent? The answer is "yes," and it's important to understand the situations for which it's not appropriate to add the type of intelligence and connectivity that we've been discussing. In general, adding intelligence may not make sense under one or more of the following conditions:

- the additional product development and manufacturing costs cannot be recouped from the customers within a reasonable time frame,
- the end user is either unable or unwilling to supply the infrastructure required to power and/or communicate with the intelligent device, or
- the physical constraints of a particular application preclude adding the additional circuitry required to implement the intelligence and connectivity.

Development and Production Costs Exceed Customer-valued Benefits

In order for any product to remain viable over the long term, customers must be willing to pay enough for the device to cover the cost to develop and manufacture it. That principle holds just as true for leading-edge technical products such as intelligent sensors as it does for more prosaic products like paper towels; no company can long afford to make a product for which it receives less money than it costs to make. Before investing the time and resources to add intelligence to its devices, a sensor manufacturer needs to determine whether its customers will be willing to pay enough of a differential in price or services to cover at least the cost of development and any increased production expenses (less any savings the manufacturer may enjoy based on the new design). Unless customers sufficiently value the benefits that an intelligent device offers, the manufacturer is better off sticking to nonintelligent products.

At first, it might seem that customers would clearly see the benefits of adding intelligence, but some applications are so cost-driven and have such razor-thin margins that customers are completely unwilling to invest in new technology. An example of this would be low-end disposable plastic cutlery, a commodity for which manufacturers receive a fraction of a cent of profit per finished item. With such miniscule profit margins, producers of this type of product simply will not spend much money on equipment; their buying decisions are focused on the bottom-line purchase price, and anything that even appears to be optional holds no value at all.

Lack of Necessary Infrastructure

A second major condition under which intelligent sensors should not (or cannot) be used occurs when a customer lacks the minimum level of infrastructure required to support both the sensors' power requirements and their communication channels. Closely related to the previous condition, in which the sensor manufacturer couldn't cost-justify building the products, this condition is one in which the customer is unable to economically justify adding the additional infrastructure needed in order for the sensors to work. Both power and communication channels are mandatory for intelligent connected sensors; without power, the sensors can't even turn on, and without communication channels they are unable to report their information.

This aspect of implementing intelligent sensor systems should not be underestimated. Although more and more manufacturing plants are becoming wired for digital data networks, it still represents a significant cost to the customer, one that many find to be a deal killer. Some of the new networking protocols provide power along with the wires used for communications (for example, Power-over-Ethernet (PoE)), but older plants in particular can be very expensive to wire. Newer low-power wireless sensors are coming into the market to help address these twin issues, but such solutions tend to be more expensive to purchase (although their long-term total cost of ownership may be lower).

Environmental Conditions Preclude Additional Electronic Circuitry

The final, and least common barrier to the use of intelligent sensors occurs when the environmental conditions of a particular application preclude the use of any additional electronic circuitry. Such conditions might be size, operating temperatures, severe vibration, or exposure to caustic chemicals. In these cases, a hardened standard sensor may be the only option, although the sensor's performance can often be significantly improved by converting the measured parameter to a digital value as soon as possible.

1.7 Real-world Examples of Intelligent Sensors

Before wrapping up this chapter, let's look at three examples of intelligent sensors in the real world, two that come from the industrial process-control market and one from the vehicular-control market.

Multichannel Digital Temperature Sensor

Temperature is a widely used parameter in the control of various industrial processes, and one of the most common temperature sensors is the thermocouple. In some ways, a thermocouple is an extremely simple sensing element; it consists of two dissimilar metals joined together at a single point and, due to what's known as the *Seebeck effect*, the junction of these two metals produces a voltage that varies with the temperature of the junction (see Figure 1.4).

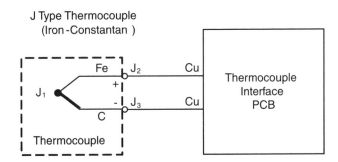

Note that there are actually three junctions of dissimilar metals and thus effectively three thermocouples in the circuit:

J_1 – sense thermocouple (the only intentional thermocouple)
J_2 – junction of Iron lead with copper PCB trace
J_3 – junction of Constantan lead with copper PCB trace

Figure 1.4. Diagram of a Basic Thermocouple

We'll examine thermocouples in detail in a later chapter, but for now the important concept to understand is that the voltage it produces is very small, on the order of millivolts, and is frequently measured in the presence of significant levels of electronic noise, which may be on the order of hundreds of volts. Complicating matters is the fact that the temperature response of thermocouples is nonlinear, so a linearization operation usually must be performed before the temperature reading can be used. There are other serious challenges in using thermocouples, but these two are sufficient to illustrate how intelligent sensors can overcome these issues to provide accurate readings in an extreme environment: an injection-molding machine.

For those readers unfamiliar with injection molding, it is a manufacturing process in which solid plastic pellets are heated to between 300°F and 900°F to melt them. The melted plastic is then injected into a mold under high pressure (on the order of 10,000–30,000 psi), and the plastic is then allowed to cool back to a solid in the shape of the mold. This process is repeated rapidly so that the manufacturer can make parts as quickly as possible. The key to running a successful injection-molding operation is to keep cycle times (the time it takes to open and close the mold once) short and scrap rates low. So long as a molder can produce good quality parts at a profit per part, he essentially has the ability to "print money" based on the speed at which he can run his cycle. An important aspect is the proper regulation of the temperature of the plastic at various points throughout the molding machine, which requires the distribution of temperature sensors (thermocouples) at key points in the process.

Unfortunately, one of the drawbacks to using thermocouples is that the wire used to create them is expensive. Molding machines in general are not particularly small, and the machines employ multiple zones of temperature monitoring and control (250 zones or more in the larger systems). The thermocouple wires thus must be run long distances to their associated temperature controllers, resulting in the worst of all possible worlds: multiple strands of expensive wire that have to be run long distances. One pioneering company in the temperature-control field realized that they could save their customers a tremendous amount of money by digitizing the temperature readings at the mold itself and then shipping the digitized readings to the controller via standard (and inexpensive) copper cables. Furthermore, they could do this for many channels of thermocouple readings and, since the thermocouple readings changed relatively slowly, the digitized readings could be time-multiplexed when sent to the controller. In the end, up to 96 channels of thermocouple data could be reported for each device, thus turning a costly, noise-prone system of long thermocouple wires into an easily managed single pair of copper wires.

Flow Sensors

While the plastic must be kept in a molten state until it gets to the mold, once there the goal is to solidify the molten plastic in the desired shape. To do this, cooling channels are built into the molds that circulate cool water or other fluids to remove the heat from the plastic quickly. If the flow of the coolant fluid is impeded, the coolant will warm up because it is staying in contact with the hot mold longer. This in turn reduces the cooling efficiency of the coolant and lengthens the time it takes for the part to solidify, thus lengthening the injection cycle time and killing profitability. Since this is obviously not something any rational molder wants to happen, smart molders include flow sensors in the coolant systems to ensure that the coolant is flowing within a desired range. If one knows some of the characteristics of the cooling fluid itself, can measure the temperature of the fluid at two different points, and knows the rate of flow, it is also possible to calculate the number of BTUs transferred from one point to the other in the cooling system.

Flow sensors come in a variety of configurations, but one of the most popular types is what's known as the *in-line flow sensor*. In this type of sensor, a propeller-like device is inserted in-line with the coolant flow, with the speed of the propeller indicating the flow of the fluid through the sensor. Over time, the bearing on which the propeller is situated will wear, resulting in an eccentric motion of the propeller that significantly degrades the quality of flow reading. One leading company developed a handheld portable unit that used special filtering algorithms to compensate for propeller bearing wear, which allowed accurate flow readings to be maintained longer and significantly extended the life of the sensor.

Steer-by-Wire Steering-position Sensor

The final example of a real-world intelligent sensor comes from the vehicular control market, specifically the steer-by-wire market for marine vehicles (boats). In a normal mechanical steering system, there is a physical link between the steering wheel and the steering-control surface (the wheels of a car, for instance, or the rudder of a boat). When the driver turns the wheel one way, the motion is translated through a series of mechanical linkages into the corresponding change in the steering control surface. Depending upon how the system is configured, the driver gets feedback from the steering system (either from the road or the water), which helps the driver adjust his actions accordingly. Although reliable, mechanical steering systems suffer from the inability to have more than one steering wheel without extremely complex (and expensive) mechanical fixtures. That's not such a big deal in a car, where only one steering wheel is normally needed, but it can be a problem for large boats, in which it would be very helpful to be able to have one steering wheel at the front (bow) of the boat when docking and one at the rear (stern) of the boat during normal cruising.

In a steer-by-wire system, by contrast, most of the mechanical linkages are replaced by electronic controls; although the driver may turn a steering wheel, that wheel is linked electronically, not mechanically, to the control surfaces. This offers several immediate advantages, not the least of which is a significant reduction in the size and weight required for the steering system. In addition, one can more easily accommodate two or more steering wheels since they can be linked by an electronic cable without requiring additional mechanical linkages. One drawback to steer-by-wire, however, is that until recently the driver had no feedback from the control surfaces, which could cause a disconcerting feeling of disconnection between the driver's actions and the resulting response of the vehicle. With the advent of a special material that changes its properties based on the strength of a magnetic field passed through it, that lack of feedback has changed. Using a steering sensor that measures the position of the wheel many times a second, a unique steer-by-wire system developed by a global innovator adjusts the feedback to the driver by adjusting the density of the special material based on a number of factors, including how quickly the driver is turning the wheel. In addition, the feedback to the driver can be changed based on conditions on the control surface, giving the driver not only a more enjoyable driving experience, but also a safer one.

1.8 Outline of the Remainder of the Book

Now that we have a basic understanding of what intelligent sensors are, have looked at how they're changing our world, and have seen some real-world examples of intelligent sensors in action, we'll conclude this chapter with a look at what's coming in the rest of the book.

Although most chapters of the book can be read in isolation, it's recommended that Chapter 2, *Intuitive Digital Signal Processing*, be read before Chapter 3, *Underneath the Hood of the dsPIC DSC*, because several of the concepts discussed in Chapter 2 are needed to fully understand the value of certain functionality in the dsPIC DSC. In Chapter 2, we develop an intuitive understanding of the digital signal processing (DSP) principles that we'll need to implement the three types of intelligent sensors that we'll explore in depth later in the book. The goal is not to turn the reader into a DSP guru; there are plenty of excellent books out there that treat the subject far more thoroughly than we do here. Instead, the intent is to develop in the reader an understanding of why we apply certain DSP principles so that the reader gains a firm grasp of when to use a particular approach and the conditions under which another is likely to be more successful.

Building on the DSP foundation developed in Chapter 2, Chapter 3, *Underneath the Hood of the dsPIC DSC*, examines in detail the dsPIC DSC, a family of digital signal controllers that is remarkably well-suited to the implementation of intelligent sensors. We will focus our attention on a particular part, the dsPIC30F6014A, which contains many features common to all dsPIC chips. Of necessity, we'll also look at the software development environment for the dsPIC DSC, as well as a number of software tools and libraries that Microchip provides to speed product development when using the dsPIC chips. At the conclusion of Chapter 3, we'll explore a basic sensor software framework that we'll use to build specific intelligent sensors in later chapters.

Chapter 4, *Learning to Be a Good Communicator*, furthers our understanding of the dsPIC device by examining the important subject of communication between the processor and other devices. These devices may be off-chip but on-board peripheral components that enhance system functionality, or they may be entire other systems that are physically distant from the sensor unit itself. As one might expect, the disparate natures of these two situations require differing communication techniques. Fortunately, the dsPIC device is well-endowed with a variety of communication interfaces that meet these challenges well. In particular, we'll look at the communications using the RS232/485 protocol, the Serial Peripheral Interface (SPI), and the Control Area Network (CAN). These are all industry-standard communication interfaces, and the ability to use them allows dsPIC-based devices to converse with a wide variety of systems in a broad universe of operating environments. In particular, CAN provides robust medium-speed (up to 1 Mbps) communications between multiple nodes on a network and is found in a huge number of industrial and vehicular products. As with many communication protocols, having a guide to sort out what's important and what's not will save the reader significant time, effort, and frustration.

Chapter 5, *A Basic DSP Toolkit for the dsPIC DSC*, continues the work begun in Chapters 3 and 4 by assembling a toolkit of modular software components that we'll use to perform specific tasks on the dsPIC processor. Once we've put together our toolkit, we'll conclude the chapter by implementing a multichannel filter bank on top of the sensor software framework introduced in Chapter 3, which will allow us to filter (remove the noise) from several channels of signal data reliably and quickly.

Moving into Chapters 6 through 8, we'll explore in depth three specific applications of intelligent sensors: temperature, pressure, and flow. These chapters will continue to build on the sensor software framework developed in Chapters 3 and 4 and the DSP building blocks developed in Chapter 5 to implement robust intelligent sensors for each particular application. By the time we've finished Chapter 8, we will have a thorough grounding in the design of intelligent sensors and how to implement them using the dsPIC DSC.

The main section of the book concludes with Chapter 9, *Where Are We Headed?*, which looks at the future of intelligent sensors and the application of digital signal controllers. Including a discussion of the merits of open vs. proprietary protocols and of the trend toward ubiquitous networking of devices, the chapter is intended to make readers aware of the key issues they should consider when designing products for the future.

In addition to the main portion of the book, there are three appendices that contain information that seems to stand best on its own, leaving to the reader the decision whether to wade through it for greater insight or to simply leave it until required. Appendix A describes the software that is provided on the CD accompanying the book, while Appendix B is devoted to the initialization of the dsPIC device and the associated system start-up code. In Appendix C, we address the challenging but oft-ignored aspects of system operation, namely interrupt-driven buffered serial I/O, which can be used for debugging or regular communications. Since no single book can rightfully claim to be the *all*-encompassing book on either intelligent sensors or the dsPIC DSC, this book includes an HTML file on the CD that contains a list of other sources the reader can use to learn more about these subjects.

Finally, for the most up-to-date information on this book, as well as additional sample programs and resources, please go the website www.intelligentsensordesign.com.

Endnotes

1. *Management Challenges for the 21ˢᵗ Century*, by Peter F. Drucker. Collins, 2001.

2. According to the CIA World Factbook, the estimated total world population as of July 2005 was 6,446,131,400. http://www.cia.gov/cia/publications/factbook/rankorder/2119rank.html

3. Based on a study (GB-200N Industrial Sensor Technologies and Markets) by B. L. Gupta for Business Communications Company, Inc. in which the 2004 industrial sensor market size in the United States was $6.1 B, with an anticipated annual growth rate of 4.6%. http://www.bccresearch.com/instrum/GB200N.html

4. Ohm's Law is $V = I * R$, where V is the voltage measured across a resistance (in volts), I is the current through the resistance (in amps), and R is the value of the resistance itself (in ohms). Ohm's Law holds true for a purely resistive element, which is all we're worried about in this example.

5. Branch intensive software is software that makes frequent changes, known as *branches*, in the processing of its program instructions. *Computationally intensive* software is software in which a significant portion of the processing time is devoted to performing mathematical computations.

6. *Four Strategies for the Age of Smart Devices*, by Glen Allmendinger and Ralph Lombreglia. Harvard Business Review, October, 2005. Reprint R0510J.

2

Intuitive Digital Signal Processing

If you have an idea, that is good. If you also have ideas as to how to work it out, that is better.

—*Henry Ford*

To this point, we've seen how valuable intelligent sensors can be to both end users and those who manufacture and sell them. It's now time to delve more deeply into what it takes to make intelligent sensors work. The first step in that journey is to develop a solid, intuitive understanding of the principles of digital signal processing, or DSP. Unlike many introductory DSP texts, the focus here will be on presenting and using the important concepts rather than deriving them, for the simple reason that addressing the subject in depth is a book-sized, not a chapter-sized, project. Other authors have already done an excellent job of addressing the topic in a more rigorous manner,[1] and our goal here is not to try to condense their work to meaningless bullet points but rather to understand how to use certain key concepts to turn raw sensor data into meaningful sensor *information*. By the end of this chapter, the reader should be comfortable identifying the key signal processing requirements for typical applications and be able to determine the appropriate process for extracting the desired measurements.

2.1 Foundational Concepts for Signal Processing

Although this discussion of DSP isn't as rigorous as most academic treatments of the subject, it's essential that we establish a clear understanding of several key concepts that form its foundation. Beginning with precise definitions of what we mean when we refer to "signals" and "noise," the discussion moves into the analysis of signals in both the time and frequency domains and concludes with an introduction to filtering, a technique that is commonly used to extract the desired information from noisy data. Section 2.2, *Applications of DSP Theory*, builds on this foundation as we begin to create a generalized framework that we can apply to a wide variety of intelligent sensing applications.

Getting Specific—What We Mean by Signals and Noise

The dictionary defines the term "signal" as "an impulse or fluctuating electric quantity, as voltage or current, whose variations represent coded information,"[2] and this definition serves well as a starting point. One interesting characteristic of electronic signals is that they operate under the principle of superposition. This principle states that the value of two or more signals passing through the same point in the same medium at a particular point in time is simply the sum of the values of the individual signals at that point in time. For example, if we had N different signals denoted $V_0(t)$, $V_1(t)$, ..., $V_{N-1}(t)$, the resulting signal $V_S(t)$ that is the superposition of the N signals would be represented mathematically as:

$$V_S(t) = V_0(t) + V_1(t) + \ldots + V_{N-1}(t)$$

Equation 2.1

It turns out that the principle of superposition is a very powerful tool; using it, we can often deconstruct complex sensor signals into separate, more basic components, which may simplify the analysis of the problem and the design of the resulting system. The real-world sensor examples developed later in this book make extensive use of this principle in the creation of the appropriate signal-processing techniques for each specific application, but first let's examine one way in which superposition leads to a better understanding of all sensor signals.

Consider the circuit shown in Figure 2.1a, which contains a thermocouple connected to a voltmeter in an idealized environment. As discussed in Chapter 1, the thermocouple produces an analog output voltage $V_T(t)$ that varies over time t with the temperature of the thermocouple junction. In this case, the measured signal $V_M(t)$ is simply the "true" signal $V_T(t)$, and the information coded in it is the temperature of the thermocouple junction.

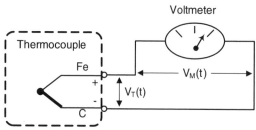

Figure 2.1a. Basic Idealized Thermocouple Circuit

Unfortunately, such an idealized environment exists only in our imaginations, much like a perfectly silent library exists only in a librarian's fantasy. Just as even

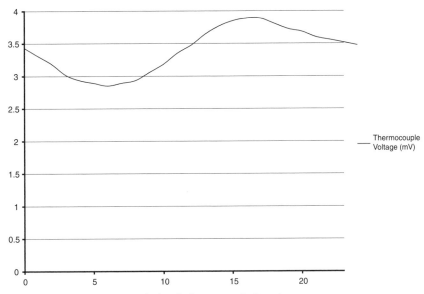

Figure 2.1b. Example of an Idealized Thermocouple Signal

the quietest library has some audible noise, real-world circuitry contains electronic noise that comes from both the surrounding environment and the components used to create the circuit. Thus, a more accurate representation of the basic thermocouple circuit would include an electrical noise generator that produces a noise voltage component $V_N(t)$ superimposed on the "true" thermocouple signal $V_T(t)$, as is shown in Figure 2.3a.

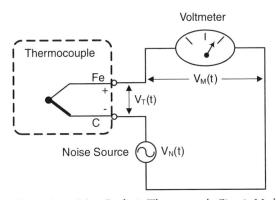

Figure 2.2a. More Realistic Thermocouple Circuit Model with Noise

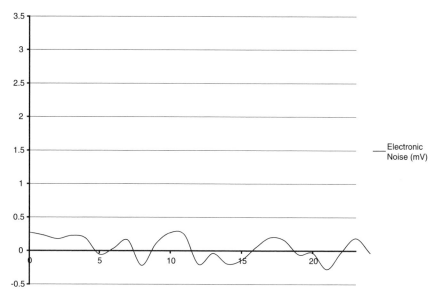

Figure 2.2b. Example of an Electronic Noise Signal

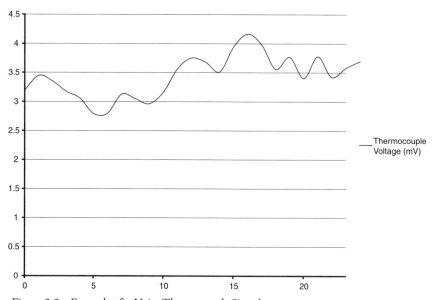

Figure 2.2c. Example of a Noisy Thermocouple Signal

To an outside observer, this distinction is not actually discernible; they simply see the measured voltage $V_M(t)$ that contains both components and is equal to:

$$V_M(t) = V_T(t) + V_N(t)$$

Equation 2.2

Of course, the end user generally doesn't want the noise component to be a part of the measured signal; after all, he's interested in the "true" thermocouple signal that has the information of interest, not a corrupted signal that distorts that information. Depending upon the characteristics of the signal of interest and of the noise, it can be possible to accurately extract the signal of interest even in the presence of significant levels of noise using the techniques to which we will now turn.

Viewing Signals in the Frequency Domain

Any real analog[3] signal can be represented in the frequency domain via a mathematical operation known as the *Fourier transform*, and the proper choice of domain (either the time domain, which is what we measure using an oscilloscope or voltmeter, or the frequency domain) can greatly simplify the analysis of a particular signal-processing situation. The basic premise of the Fourier transform is that continuous, linear[4] time-domain signals (like the voltages we're measuring in the examples above) can be accurately represented by the superposition of orthogonal sinusoidal signals of varying frequencies. That's a tremendous amount of fairly technical mathematical jargon, but what's valuable about this operation is that it allows us to fairly easily determine the frequencies in which most of the energy of the signal occurs, which is essentially telling us what the most important parts of the signal are.

An example may help clarify the point. Consider the purely sinusoidal signal in Figure 2.3a, and its frequency-domain counterpart in Figure 2.3b. Note the very interesting relationship between the two domains: a continuous signal in the time domain actually maps to two spikes in the frequency domain symmetrically distributed along the frequency axis! Although the graph in Figure 2.3b shows two peaks, the spread of the frequency spectrum is due to limitations of the discrete mathematics used by the program that generated the image. Mathematically, in the frequency domain there are two spikes located exactly on the sinusoid's frequency. If we construct a more complex signal by adding a second sinusoid to the first, the principle of superposition tells us that we might get a new signal that looks like that shown in Figures 2.4a and 2.4b. Here we see a standard DTMF (dual tone multifrequency) signal, just as you might get if you punched a digit on your touchtone phone. The addition of a single extra frequency has caused the signal to lose much of its sinusoidal appearance in the time domain, but the same signal in the frequency domain is simply four spikes, with the two additional spikes corresponding to the new frequency.

Example Sinusoidal Signal

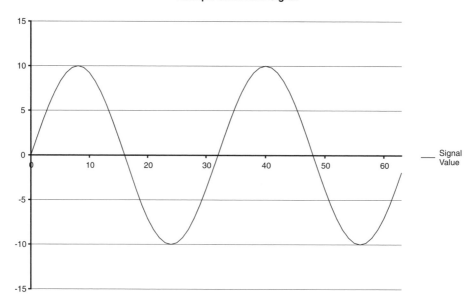

Figure 2.3a. Time-Domain Sinusoidal Signal

Magnitude of Sinusoidal Signal's Frequency Domain Representation

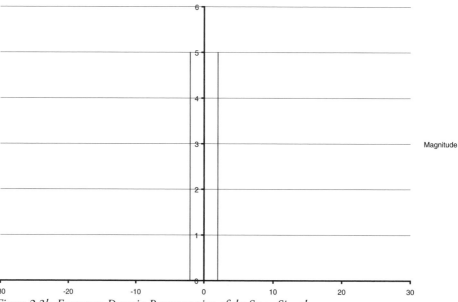

Figure 2.3b. Frequency-Domain Representation of the Same Signal

DTMF Signal for the "5" Key

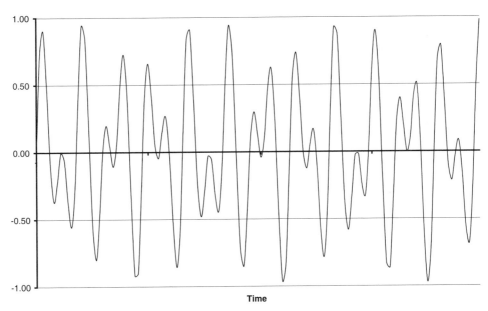

Figure 2.4a. DTMF Time-domain Signal

Magnitude of DTMF Frequency Component

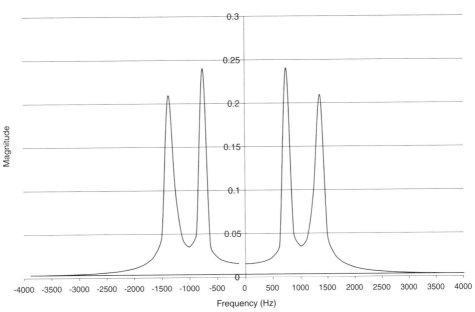

Figure 2.4b. DTMF Frequency-domain Representation

Figures 2.4a and 2.4b demonstrate a very powerful aspect of the Fourier transform: the superposition principle holds in both the time and the frequency domains. Signals that are added together in the time domain have a frequency spectrum that is the sum of the spectra[5] of the individual signal components. Simply by viewing this signal in the frequency domain, the designer can rapidly identify its constituent parts, which will be of great use in analyzing and designing the processing required to extract the information of interest. This concept of *spectral analysis*, the analysis of the frequency domain representation of a signal, is a powerful one that we will apply in the real-world examples in later chapters.

Two terms that often arise when performing spectral analysis are *frequency band*, which simply means a continuous range of frequencies, and *bandwidth*, which generally refers to the highest frequency component in a signal. For example, in Figure 2.4b, the designer might be interested in the frequency band from 770 Hz to 1477 Hz, which contains the two frequencies that make up that particular DTMF signal. Since 1477 Hz is the highest frequency signal component, the theoretical bandwidth for the DTMF signal is 1477 Hz.

There is one additional aspect to the time-domain–frequency-domain representation issue that is important to understand, and that is the fact that rapidly changing signals in the time domain generate a broader corresponding spectrum in the frequency domain, while slowly changing signals produce a narrower overall frequency spectrum limited to lower frequencies. As we'll see in the next section, sensor designers can use this fact to determine the optimal approaches to removing noise from the signals of interest. Figures 2.5a, 2.5b, and 2.5c show frequency domain representations of low-frequency (slowly changing), high-frequency (rapidly changing), and broadband (low- and high-frequency) signals. In practice, signal distortion will spread the actual DTMF bandwidth somewhat beyond the 1477 Hz theoretical value.

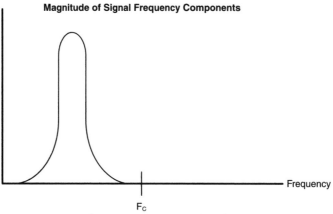

Figure 2.5a. Low-frequency Content Signal in the Frequency Domain

Magnitude of Signal Frequency Components

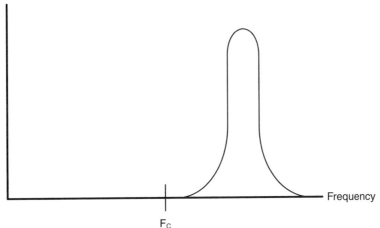

F_C

Figure 2.5b. High-frequency Content Signal in the Frequency Domain

Magnitude of Signal Frequency Components

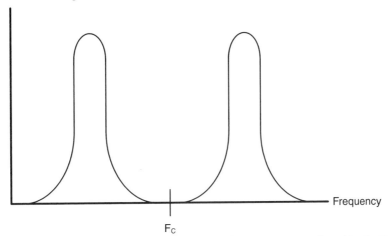

F_C

Figure 2.5c. Combination of Low- and High-frequency Content Signal in the Frequency Domain

Cleaning Up the Signal—Introducing Filters

We're all familiar with the general idea of a filter: it removes something that we don't want from something we do want. Coffee filters that pass the liquid coffee but retain the grounds or air filters that pass clean air but trap the dust and other pollutants are two common examples of mechanical filters in everyday life. That same concept can be applied to noisy electrical signals to pass through the "true" signal of interest while blocking the undesirable noise signal.

Looking at Figure 2.5c, imagine for a moment that the signal of interest is in the lower-frequency region and that the noise signal is in the higher-frequency region. Ideally, we'd like to be able to get rid of that high-frequency noise, leaving just the signal component that we want. We can picture the process that we'd like to perform as one in which we apply a mask in the frequency domain that passes all of the low-frequency signal components without affecting them at all but that zeros out all of the high-frequency noise components. Graphically, such a mask might look like the frequency spectrum shown in Figure 2.6a. If we multiply each point in the graph of Figure 2.5c by the corresponding point in the graph of the mask in Figure 2.6a, we get the resulting frequency spectrum shown in Figure 2.6b, which is precisely what we want.

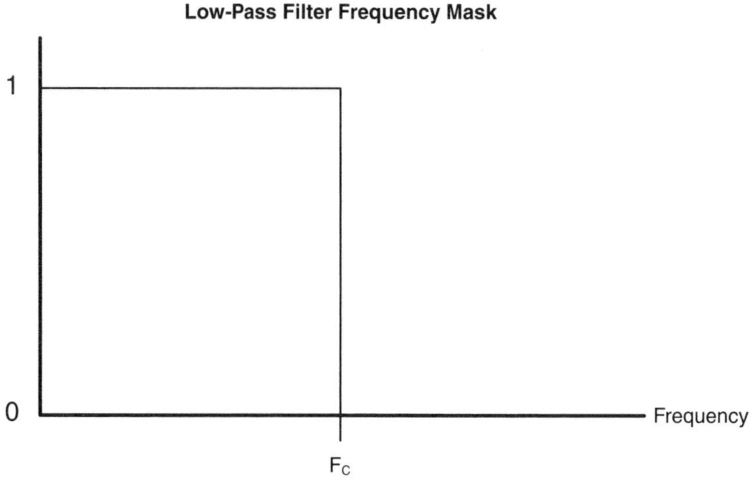

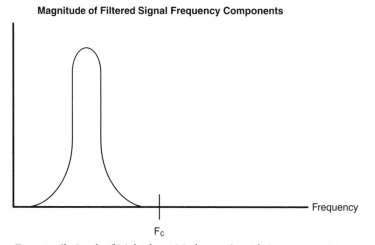

Figure 2.6a. Example Frequency Mask

Figure 2.6b. Result of Multiplying Mask in 2.6a with Spectrum in 2.5c

Thought experiments like these are helpful, but is it possible to implement this in the real world? The answer is "yes," albeit with some important qualifications that arise from deviations between real-world and idealized system behavior. Before we get into those qualifications, though, let's take a look at an important foundational concept: sampling.

Sampling the Analog Signal

Sensor signals are inherently analog signals, which is to say that they are continuous in time and continuous in their value. Unfortunately, processing analog signals as analog signals requires special electronic circuitry that is often difficult to design, expensive, and prone to operational drift over time as the components age and their properties change. A far better approach is to convert the input analog signals to a digital value that then can be manipulated by a microprocessor. This technique is known as *analog-to-digital conversion*, or *sampling*.

Figure 2.7a shows an example of a continuous time voltage signal, and Figure 2.7b shows the sampled version of that signal. One key concept that can sometimes be confusing to those who are new to sampled signals is that the sampled signal is simply a sequence of numeric values, with each numeric value corresponding to the level of the continuous signal at a specific time. For a sampled signal such as that shown in Figure 2.7b, the signal is only valid at the sample time. It is not zero-valued

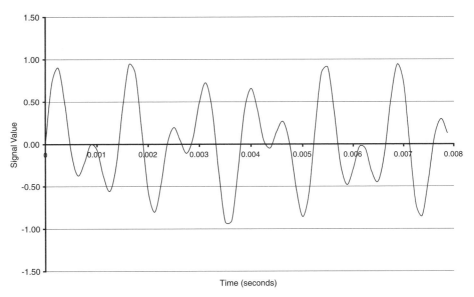

Figure 2.7a. Example of a Continuous-time Voltage Signal

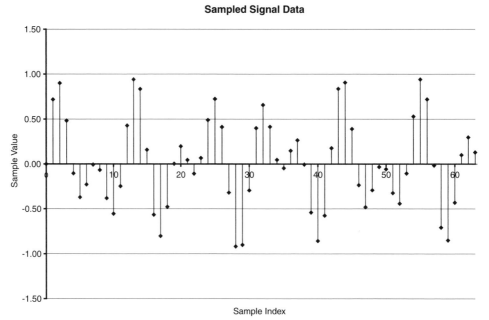

Figure 2.7b. Corresponding Sampled Version of the Signal in Figure 2.7a

between samples, but the convention for presenting sampled data graphically is to display the sample values on a line (or grid), with the X-axis denoting the parameter used to determine when the data is sampled (typically time or a spatial distance).

Another convention is to associate sampled signal values in a sequence using an *index notation*. In this scheme, the first sample of the signal $x(t)$ would be x_0, the second sample would be x_1, and so on. If we add two signals x and y, then the resulting signal z is simply the sample-by-sample addition of the two signals:

$$z_0 = x_0 + y_0$$

$$z_1 = x_1 + y_1$$

.

.

$$z_N = x_N + y_N$$

Sampling has two important effects on the signal. The first of these effects is what's known as *spectral replication*, which simply means that a sampled signal's frequency spectrum is repeated in the frequency domain on a periodic basis, with the period being equal to the sampling frequency. Figures 2.8a and 2.8b show an example of the frequency spectrum of an example signal and the resulting frequency spectrum of the sampled version of the signal.

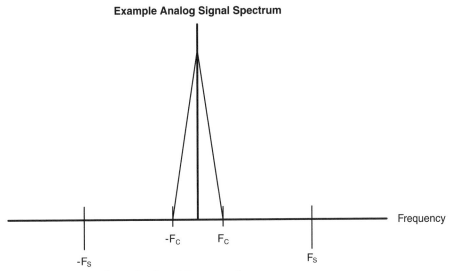

Figure 2.8a. Example Analog Signal Frequency Spectrum

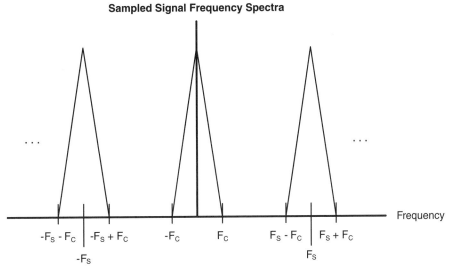

Figure 2.8b. Corresponding Frequency Spectrum of the Sampled Signal

As one can easily see, a problem arises when the highest frequency component in the original signal is greater than twice the sampling frequency, a sample rate known as the *Nyquist rate*. In this case, frequency components from the replicated spectra overlap, a condition known as *aliasing* since some of the higher frequency components in one spectrum are indistinguishable from some of the lower frequency components in the next higher replicated spectrum. Aliasing is generally a bad condition to have in a system and, although the real world precludes eliminating it entirely, it is certainly possible to reduce its effects to a negligible level.

Let's look at a simple example to illustrate how aliasing can fool us into thinking that a signal behaves in one way when in reality it behaves totally differently. Imagine that we are sampling the position of the sun at various times during the day over an extended period of time. Being good scientists, we want to verify that our sampling rate really does make a difference, so we decide to take two sets of measurements using two different sampling rates. The results from the first set of measurements, which employ a sampling rate of once every 1.5 hours, are shown in Figure 2.9a. As we would expect, the measurements show that the sun proceeded from east to west during the course of the experiment.

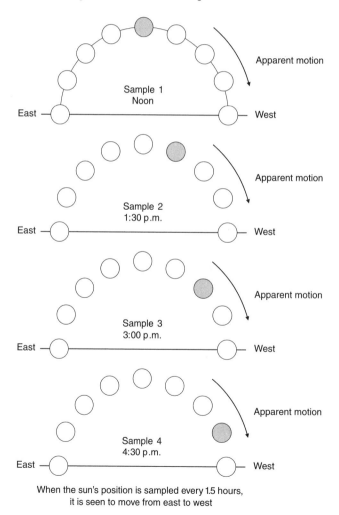

When the sun's position is sampled every 1.5 hours,
it is seen to move from east to west

Figure 2.9a. Sun's Position Sampled Every 1.5 Hours

Now take a look at the results from the second set of measurements, which have a sampling rate of once every 22.5 hours. From the data, we can see that the sun appears to move from west to east, just the opposite of what we know to be true! This is exactly the type of error one would expect with aliasing, namely that the signal characteristics appear to be something other than what they really are (hence the term *aliasing*).

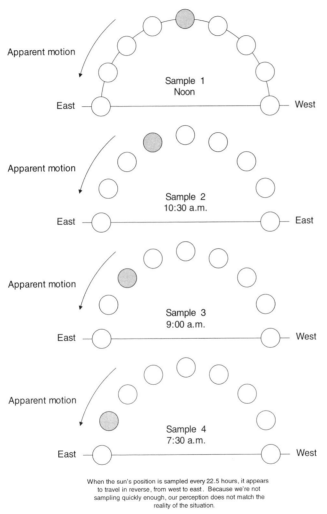

When the sun's position is sampled every 22.5 hours, it appears to travel in reverse, from west to east. Because we're not sampling quickly enough, our perception does not match the reality of the situation.

Figure 2.9b. Sun's Position Sampled Every 22.5 Hours

Low-pass Filters

We've seen an example of the first type of filter, the low-pass filter, which passes low-frequency components and blocks high-frequency signal components. An idealized example of a low-pass filter is shown in Figure 2.10, in which the passband, the frequency range of the signal components that are passed, is 1500 Hz wide. Note that the bandwidth in this case is also 1500 Hz, since that's the highest frequency component of the filter.

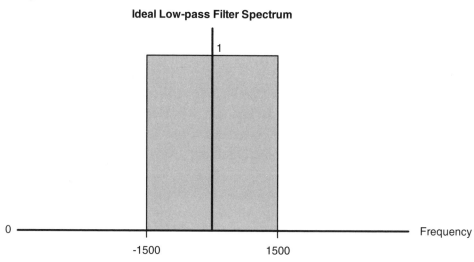

Figure 2.10. Idealized Low-pass Filter with a Bandwidth of 1500 Hz

Low-pass filters are probably the most widely used type of filter for the simple reason that, in the real world, we don't deal with signals of infinite bandwidth. At some point, the frequency content of a signal drops off to insignificance, so one of the most common approaches to noise reduction is to establish some limit for the frequency components that are considered to be valid and to cut off any frequencies above that limit. For example, when we are using thermocouples to measure temperature, the thermocouple voltage can change only so quickly and no faster because the temperature of the physical body that is being monitored has a finite rate at which it can change (i.e., the temperature can't change discontinuously). In practice, this means that the frequency components of the temperature signal have an upper bound, beyond which there is no significant energy in the signal. If we design a low-pass filter that will cancel all frequencies higher than the upper bound, we know that it must be killing only noise since there are no valid temperature signal components above that cutoff frequency.

High-pass Filters

A complement to the low-pass filter is the *high-pass filter*, which passes only high-frequency signal components and blocks the low-frequency ones. In the idealized high-pass filter of Figure 2.11, the passband starts at 1500 Hz and continues to all higher frequencies. Note that the bandwidth in this case is infinite since all frequencies starting with the passband are included in the filter.

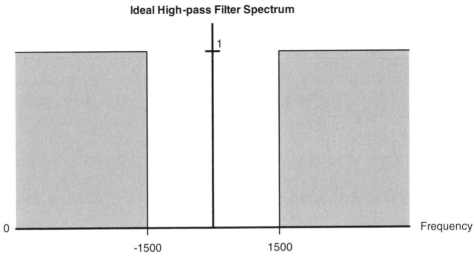

Figure 2.11. Idealized High-pass Filter with Passband Starting at 1500 Hz

Since we just stated that no real-world signal has infinite bandwidth, why would we want to use a filter that seems to assume that condition? In some cases, the signal we're measuring is an inherently AC signal; by the nature of the system anything below a certain frequency is obviously noise because no valid signal components exist below that frequency. An example of this might be the auditory response of the human ear, which is sensitive only to frequencies in the range of 20 Hz to about 20 kHz. Anything below 20 Hz is of no practical value and can be treated as noise.

Bandpass Filters

A bandpass filter is essentially the combination of a high-pass filter and a low-pass filter in which the passband of the high-pass filter starts at a lower frequency than the bandwidth of the low-pass filter, as shown in Figure 2.12. Here we see that the filter will pass frequencies between 750 Hz and 1500 Hz while blocking all others.

Bandpass filters are used whenever the designer wants to look at only a particular frequency range. A very common example of this is the tuner in a radio, in which the tuner uses a bandpass filter with a very narrow passband to isolate the signal from an individual radio station. With the tuner, the goal is to pass the signal from the station of interest as clearly as possible while simultaneously attenuating the signals of all other stations (presumably at lower or higher frequencies) to the point where they are inaudible.

Bandpass filters are also commonly used to look at the strength of the signal in certain passbands of interest. DTMF detectors use this principle to determine what key a person has pressed on their touchtone phone. In a DTMF system, each

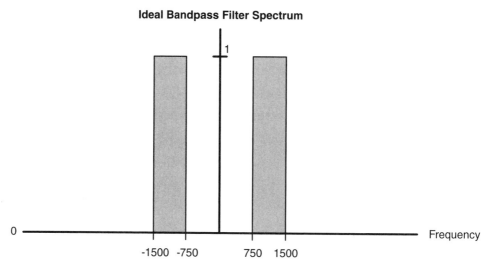

Figure 2.12. Idealized Bandpass Filter

key is represented by a combination of two and only two frequencies that have no common harmonics. These two frequencies always have one component from a group of four low-frequency values and a second component from a group of four high-frequency values, as is shown in Table 2.1.

	1209 Hz	1336 Hz	1477 Hz	1633 Hz
697 Hz	1	2	3	A
770 Hz	4	5	6	B
852 Hz	7	8	9	C
941 Hz	*	0	#	D

Table 2.1. DTMF Tone Combinations

Basically, to be a valid DTMF tone, each of the two frequency components needs to be within about 1.5% of their nominal value, and the difference in signal strength between the two components (known as "twist") must be less than 3 dB. Using a bandpass filter for each of the eight frequency components plus one for the overall signal bandwidth, a detector can examine the outputs of each filter to determine that only two frequency components are active at any given time, that the two components are a valid combination (one from the low-frequency group and one from the high-frequency group), and that their relative strength is acceptable.

Bandstop Filters

The bandstop, or notch, filter can be viewed as the complement to the bandpass filter in much the same way that the high-pass filter is the complement of the low-pass filter. Whereas bandpass filters allow only a relatively narrow band of frequencies to pass, bandstop filters sharply attenuate a narrow band of frequencies and leave the rest relatively untouched. Figure 2.13 shows an example of a bandstop filters.

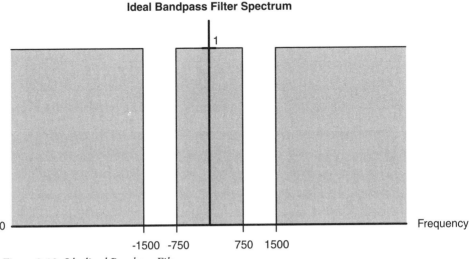

Ideal Bandpass Filter Spectrum

Figure 2.13. Idealized Bandstop Filter

By far the greatest application of bandstop filters is in the reduction of powerline noise centered around 50 Hz or 60 Hz (depending on location). In many applications, the 50-Hz or 60-Hz power signal will couple into the sensing circuitry and, unfortunately, the power signal's frequency often is in the midst of the frequency spectrum for the signal of interest. A simple low-pass or high-pass filter that would exclude all frequencies above or below the power frequency would attenuate the desired signal too much in such cases, so designers try to remove only the frequency components right around that of the power.

Digital Filter Implementations

To this point, our exploration of filters has been strictly along conceptual lines; we turn now to the actual mathematical implementation of these filters. In general, digital filters are created by applying weighting factors to one or more values of the sampled data and then summing the weighted values. For instance, if we have a sampled input signal x_i for $i = 0, 1, \ldots, N - 1$, we can generate a filtered output y_i that is given by:

$$y_i = a_0 x_0 + a_1 x_1 + \ldots + a_{N-1} x_{N-1}$$

Equation 2.3

where the a_i terms are constant weighting values that are applied to the corresponding x_i sampled input signal value. An example of a low-pass filter is an averaging filter, whose output is simply the average of a given number of samples. This smooths out the signal because noise is averaged over the entire group of samples. If we choose to average four samples to get our filtered output, the corresponding equation would be:

$$y_i = \tfrac{1}{4}\,x_0 + \tfrac{1}{4}\,x_1 + \tfrac{1}{4}\,x_2 + \tfrac{1}{4}\,x_3 \qquad\qquad \text{Equation 2.4}$$

By adjusting the weights of the individual *taps* of the filter (the sampled data values), we can adjust the filter's response. To make things easier for designers, a number of companies make digital filter design and analysis software, and free versions are available on the Internet as well. In our designs, we will use Microchip's dsPIC Filter Design™ software to create and analyze the digital filters we need.

The preceding example illustrates what is known as a *finite impulse response* or *FIR* filter structure. Filters constructed using this approach always have a fixed number of taps, and thus their output response depends only upon a limited number of input samples. If we pass the impulse signal shown in Figure 2.14 through a filter of length N taps, we know that the filter's output to the input will die out after N samples, since all subsequent input values will be zero.

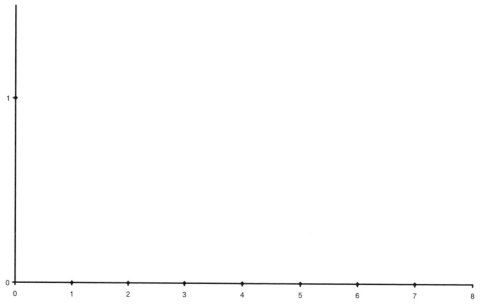

Figure 2.14. Unit Impulse Signal

Another filter structure is the *infinite impulse response* or *IIR* filter. IIR filters use both weighted input signal samples and weighted output signal samples to create the final output signal:

$$y_i = b_0 y_{i-1} + b_1 y_{i-2} + \dots + b_{i-K} y_{i-K-1} + a_0 x_0 + \dots + a_{N-1} x_{N-1} \qquad \text{Equation 2.5}$$

where the b_i terms are constant weighting terms that are applied to the corresponding y_{i-1} terms.

At first glance, it would appear that we've made the filter much more complex, but that's not necessarily the case. Looking at the four-tap averaging filter that we examined for the FIR filter, we could implement the same function as:

$$y_i = y_{i-1} + \tfrac{1}{4}\, x_0 - \tfrac{1}{4}\, x_4 \qquad\qquad \text{Equation 2.6}$$

While reducing the computational requirements by a single tap may not seem particularly important, more complicated filters can see a significant reduction in computational and memory requirements using an IIR implementation. This reduction comes at a cost, however; unlike FIR filters, IIR filters can theoretically respond to inputs forever (hence the name of the structure), which may not be at all desirable. Designers also have to be careful to ensure that errors don't accumulate or else performance can degrade to the point where the filter is unusable.

Median Filters

All of the filters that we've discussed so far are based on simple mathematical equations, so their behavior is easily analyzed using well-known and well-understood techniques. These filters tend to work best with noise that is contained to specific spectra, which is often an appropriate design model. Sometimes, however, systems are susceptible to what is called *shot* or *burst* noise, in which the measured signal has bursts of noise rather than a continuous noise signal. To counteract this, systems may employ another form of filtering known as *median filtering* that is somewhat more heuristic but does an excellent job of reducing shot noise.

In a median filter, the signal is sampled as in the other forms of filtering, but rather than performing a simple mathematical operation on the samples, the samples are ordered highest to lowest (or vice versa, it doesn't really matter which), and then the middle or *median* sample is selected. If the length of the median filter is greater than the length of the noise burst, the noisy signals should be completely eliminated. An example of a length-7 median filter and its effect upon a signal corrupted with shot noise whose burst is a maximum of three samples is shown in Figures 2.15a through 2.15d.

Sample Signal without Noise

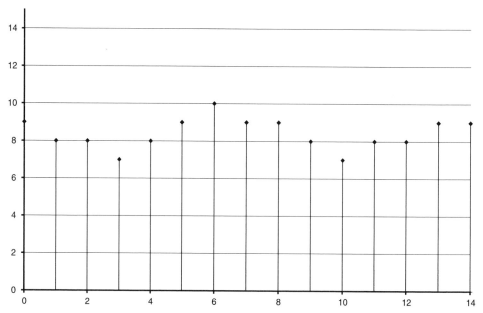

Figure 2.15a. Sample "True" Signal

Sample Shot Noise with Burst Length of 3

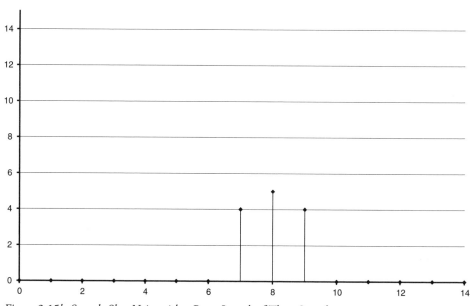

Figure 2.15b. Sample Shot Noise with a Burst Length of Three Samples

Sample Signal with Noise

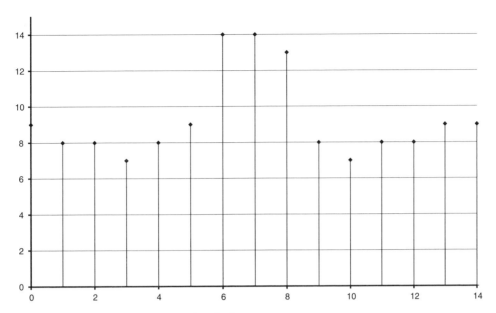

Figure 2.15c. Measured "True" Signal with Shot Noise

Signal Filtered with Length-7 Median Filter

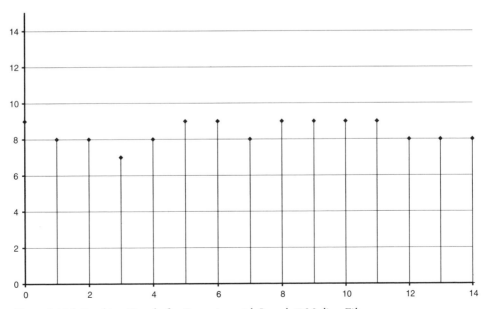

Figure 2.15d. Resulting Signal after Processing with Length-7 Median Filter

2.2   **Issues Related to Signal Sampling**

We've already touched on one problem that can arise when sampling an analog signal, namely the problem of aliasing. There are three other issues with signal sampling to which we now turn our attention: *digitization effects*, *finite register length effects*, and *oversampling*.

The Effect of Digitization on the Sampled Signal

So far, we've assumed that all of the signals we're measuring are continuous analog values—i.e., our measurements are completely accurate. Even in the cases in which we have noise, the underlying assumption is that the measurement itself, for example the noisy sensor output voltage, is known precisely. In reality, at least for a system that employs digital signal processing, that's not really true because the measured analog signals go through a process known as *digitization* that converts the analog signal to a corresponding numeric value that can be manipulated mathematically by a processor. Figure 2.16 shows this process (signal value is sampled at the points shown).

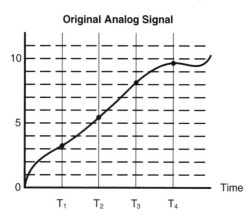

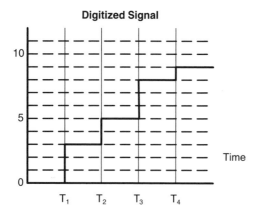

Figure 2.16. Signal Digitization Process Showing Four Successive Samples

The issue that we face with digitization is that within any processing unit we have only a finite number of bits with which to represent the measured signal. For instance, let's assume that we want to sample a signal that varies between 0V and 5V. If we try to represent the measurement with one bit, we'll have exactly two possible values (0 and 1) that we can use. Designating the measured signal voltage as V_S, we might choose to map the lower half of the signal range ($0 \leq V_S < 2.5V$) to 0 and to map the upper half ($2.5V \leq V_S < 5V$) to 1. Unfortunately, that's pretty poor resolution! While we can obviously improve the resolution significantly by using more bits to represent our numeric values, we will always map a range of input values to a particular output value, which means that almost all measured signal values within that range will be in error (the lone exception being the signal value that corresponds exactly to the numeric value).

This digitization error can be viewed as a noise signal that is superimposed on the true value of the measured signal as shown in Figures 2.17 and Figures 2.18. Note that, depending upon whether we perform the digitization by rounding the measured value (as in Figure 2.17) or by truncating the measured value (as in Figure 2.18), we will essentially have either a triangular noise signal (rounding) or a sawtooth noise signal (truncation). Although we can never completely eliminate the issue, we can reduce its significance by ensuring that we use a relatively large number of bits (say 16 to 32, depending on the application) to represent the numeric values in our algorithms. For instance, if we use 16-bit values, we can represent our signals with an

Error When Digitized Measurement is Rounded

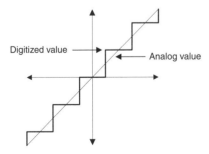

When the digitized signal is rounded, the error is evenly distributed about the "true" analog value, with a maximum absolute error of ½ of a single digitization interval

Figure 2.17. Digitization Error Introduced by Rounding

Error When Digitized Measurement is Truncated

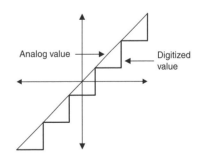

When the digitized signal is truncated, the error has the same stairstep effect, but its absolute value is now between 0 and a single digitization level

Figure 2.18. Digitization Error Introduced by Truncation

accuracy of 0.0015 % (assuming no other sources of digitization noise); using 32-bit values, that resolution improves to 2.3×10^{-8} % (since there are 2^{32} discrete levels).

Finite Register Length Effects

Closely related to digitization effects, which deal with the inaccuracy introduced by having a finite number of values available to represent a continuous signal, finite register length effects refer to the issues caused by performing repeated mathematical operations on values that are represented by a finite number of bits. The problem is that, because we have a limited number of bits with which to work, repeated mathematical operations can cause the accumulator in the processor to overflow. A simple example will illustrate the effect.

Suppose our application digitizes the input signal into a 16-bit value (0–FFFFh) and further suppose that we're using a processor with a 16-bit accumulator. If we try to average two samples that are at ¾ of the digitization range (BFFFh), we would get a value of BFFFh if we had infinite precision in the accumulator (½ (BFFFh + BFFFh) = BFFFh). However, if we add the two samples together in a 16-bit accumulator, the sum is not 17FFEh but 7FFEh since the most significant bit would be truncated (we don't have space in the accumulator for it). If we then take one-half of the sum, the average becomes 3FFFh, not the BFFFh we want.

Although we can use larger accumulators (i.e., accumulators with more bits), the problem is inherent to the system and is exacerbated when multiplication operations are included. By choosing appropriate ways to represent the numeric values internally and by carefully handling cases of overflow and underflow, designers can mitigate the effects of having finite register lengths, but the issue must always be addressed to avoid catastrophic system failures. We'll see how the dsPIC processor handles these issues in the next chapter.

Oversampling

As mentioned previously, designers have to ensure that systems that use DSP sample the input signals faster than the Nyquist rate (twice the highest frequency in the input signal) to avoid aliasing. In reality, input signals should be sampled at least four to five times the highest frequency content in the input signal to account for the differences between real-world A/D performance and the ideal. Doing so spreads the sampled spectrums further apart, minimizing bleed-over from one to the next.

Another issue with any filter is the delay between the time the input signal enters the filter and the time the filtered version leaves the filter, and this is true for digital as well as analog filters. Generally, the more heavily filtered the input signal, the greater the delay is through the system. If the delay becomes excessive (something that's application dependent), the filtered output can be worthless since it arrives too late to be of use by the rest of the system.

Oversampling, the practice of sampling the signal much faster than strictly necessary, can be employed to allow strong filtering of signals without introducing an excessive delay through the system. The oversampled signal can be heavily filtered, but since the delay is relative and the sampling rate is much higher than necessary, the filtered signal is available for use by other system components in a timely manner. The downside of this approach is that it requires greater processing power to handle the higher data rate, which generally adds to the cost of the system and its power consumption.

2.3   How to Analyze a Sensor Signal Application

When analyzing a specific sensor signal-processing application, designers need to understand the following aspects of the system:

1. the physical property to be measured,

2. the relationship between the physical property being measured and the corresponding parameter value to be reported,

3. the expected frequency spectrum of the signal of interest and of any noise sources in the environment,

4. the physical characteristics of the operating environment,

5. any error conditions that may arise and the proper technique for handling them,

6. calibration requirements,

7. user and/or system interface requirements, and

8. maintenance requirements.

Often, the natures of the physical parameter being measured and of the operating environment will help guide the designer in the selection of appropriate signal-processing capabilities to include in the sensor system. For instance, if one is measuring the temperature of a large metallic mass heated by a relatively small heating element, it's safe to assume that the frequency content of the signal of interest is minimal since the temperature can change only gradually. This means that the sensor can employ heavy filtering of the input to reduce noise. In contrast, a temperature sensor monitoring a small device being heated by a laser must be capable of reacting to intense changes in temperature that can occur very quickly. In such a situation, noise filtering must be lighter and other processing may be required to address noise that gets through the initial filters.

It's also critical to understand the relationship between the physical property being measured and the corresponding parameter being reported to the user or to the rest of the system. Does the reported parameter vary linearly with the physical property (as is the case with RTD temperature sensors), or does it have a nonlinear relationship (as do many thermocouples)? If the relationship is nonlinear, is it possible to segment the relationship into piecewise linear regions to simplify computation? A poor or incorrect understanding of the relationship between physical property and reported parameter can render a sensor system useless.

Finally, designers must always consider that sensor systems are going into the real world, where problems are guaranteed to arise at the worst times. The system must be designed to detect common errors, and the more robust its error detection and handling scheme, the better. The loss of a sensor on the production floor may stop production for an entire line, so any features that allow quick troubleshooting and easy repair are greatly appreciated by end users. Of even more importance than maintenance, however, is the ability of the sensor system to detect dangerous conditions that may lead to unsafe operation unless corrected. Sensors that operate in a fail-safe environment must be designed with rigorous attention to fault detection, reporting, and correction.

2.4   A General Sensor Signal-processing Framework

We're now ready to set up a general sensor signal-processing framework that we'll use in each of our in-depth applications in Chapters 4–7. Like all good designs, the framework is deceptively simple; the key is to implement it reliably so that it performs all of its required tasks accurately, on time, every time. The framework is shown in Figure 2.19.

The framework must be constructed as a *hard real-time* system; i.e., its response to system inputs and events must be deterministic (occur within a fixed time) and all processing for a given input or event must be finished before the next input or event occurs, at least for the critical processing sections. Less critical sections, such as the communication protocol handler, are important, but they can occur in soft real-time; they must be capable of processing all inputs or events eventually, but they can queue up those inputs or events for processing at a time that's convenient for the application.

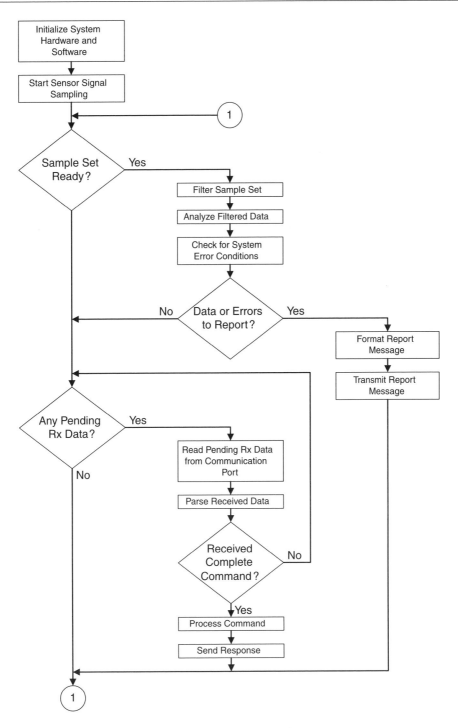

Figure 2.19. General Sensor Signal-processing Framework

Signal Conditioning and Acquisition

The signal conditioning and acquisition section is responsible for performing any required conditioning of the analog input signal to limit the frequency spectrum to a band that can be successfully processed, to amplify the signal level to an appropriate range for digitization, and to digitize the resulting analog input signal. The output of this section is a stream of sampled data that can then be processed numerically by the rest of the system.

Pre-analysis Filtering

Once the raw physical property signal has been sampled, it's often necessary to apply application-specific filtering to the signal to remove unwanted noise or to somehow shape the signal into a more useful form. The filtering is typically performed immediately after acquisition so that processing algorithms later in the signal chain are able to use relatively clean data, hopefully yielding better results.

Signal Linearization

Sometimes the parameter of interest does not vary linearly with the physical property being measured. A common example is a thermocouple signal, which has a complex polynomial relationship between its voltage and the corresponding temperature. In such cases, the signal often needs to be linearized so that it can be dealt with more easily by the parameter analysis section. The specific linearization technique employed will vary by the type of property being measured.

Parameter Analysis

The parameter analysis is also highly application-specific. Although limited only by the designer's imagination, some typical operations are parameter transformation (in which the measured signal is converted to the desired corresponding parameter value mathematically), frequency analysis, and limit comparison. Frequently, this is the most complex aspect of the sensor system and the area in which the most value can be added to the product.

Post-analysis Filtering

Once a parameter value has been computed, it's not uncommon to filter those values to smooth the data for use by other components in the system. As with the pre-analysis filtering, the particular type of filter employed is application-specific.

Error Detection and Handling

While the parameter analysis section is generally where the most unique value is added to the sensor system, the error detection and handling section can make or break the viability of the system. The ability to detect and to recover from errors can separate a product from its competition, particularly in situations in which the penalty for failure can be catastrophic. Simple error detection might include checking for the presence of the sensor element and verifying that extracted parameter values are in a reasonable range. More advanced error detection might include diagnostics to alert the user before an actual failure occurs.

Communication

The final element in the framework is the communication section. It is this section that reports all of the information gathered by an intelligent sensor and that allows the user to configure it for operation, so it is absolutely critical that this interface be robust and reliable. A wide variety of communication interfaces are available, from RS-232 to Control Area Network (CAN) to Ethernet to wireless, though not all systems support all interfaces. The designer must select an interface that provides the easiest integration of the product with other elements of the system while staying within the cost and reliability constraints necessary for a particular application.

2.5   Summary

In this chapter, we've introduced the basic concepts of digital signal processing on a conceptual level. The reader should recognize that a thorough knowledge of DSP is invaluable to the development of robust sensor systems, and this treatment has been meant to instill an intuitive, not exhaustive, understanding. Nevertheless, with this understanding we have been able to develop a general framework for the digital analysis and reporting of sensor information, one that will be used in subsequent chapters to design sensor systems for specific applications.

Endnotes

1. Two outstanding books on DSP are *Digital Signal Processing: A Practical Guide for Engineers and Scientists*, by Steven Smith and *Understanding Digital Signal Processing*, 2nd edition, by Richard Lyons.

2. *Webster's II New Riverside Dictionary*, 1984, Houghton Mifflin Company.

3. For our purposes, an analog signal is a continuous-time signal, i.e., one that is not sampled.

4. In this case, linear refers not to a straight-line function but rather to a function that obeys the principle of superposition. Mathematically, if the function is a transformation $H[]$, and $y_0(t)$ represents the response of the function to the input $x_0(t)$ while $y_1(t)$ is the response of the function to the input $x_1(t)$, then $H[]$ is linear if and only if

$$H[ax_0(t) + bx_1(t)] = aH[x_0(t)] + bH[x_1(t)] = ay_0(t) + by_1(t).$$

5 The plural of spectrum is spectra, although you may also see the term written as *spectrums*.

Underneath the Hood of the dsPIC DSC

Success is the sum of detail. It might perhaps be pleasing to imagine one's self beyond detail and engaged only in great things, but…if one attends only to great things and lets the little things pass, the great things become little—that is, the business shrinks.
—Henry Firestone

When one goes on a trip, the choice of transportation depends primarily on the purpose of the excursion and what's required to accomplish that purpose. While theoretically one could use a sports car to deliver thousands of computers to a Wal-Mart distribution center, it would be far more efficient to use a tractor-trailer to accomplish the task. The same holds true when it comes to implementing intelligent sensors; we're far more likely to achieve the required performance if we use a hardware platform that's specifically designed to support the tasks that we need to accomplish. Such a platform must offer deterministic[1] acquisition, filtering, and analysis of the signals being monitored as well as reliably handling all communications with the outside world. In many applications, the system is required to do this for multiple signals and possibly several communication channels, further increasing the needed platform processing performance.

Fortunately, a new class of processor known as the *digital signal controller* or DSC has been developed recently that marries the powerful mathematical processing performance of a pure digital signal processor with the highly deterministic behavior of standard microcontrollers. One such DSC is the dsPIC® DSC from Microchip, which integrates a tremendous amount of functionality into a single chip, allowing designers to create robust sensing and control solutions in a very small package. We'll use the dsPIC DSC in subsequent chapters to craft solutions to a variety of common intelligent-sensing applications, but first we'll examine the resources available to users of the chip. Throughout the exploration, the goal is twofold: to learn what's available and to understand why the dsPIC designers made the implementation choices that they did so that we can optimize our system to best use the chip's resources.

The dsPIC DSC can be viewed in a number of different ways, but three particularly useful perspectives are an examination of the chip's data-processing architecture, a study of the mathematical representations and operations that the chip supports, and an analysis of the various on-chip peripheral components. A thorough understanding of these three subjects will serve as a solid foundation for creating meaningful systems using the chip.

It's somewhat incorrect to talk about "the" dsPIC chip, since the dsPIC DSC is actually two related families of chips with a variety of configurations that allow the designer to select the combination of peripherals that best suits the particular application. To ground the discussion in the real world, therefore, we will focus on one specific chip, the dsPIC30F6014A, which has virtually all of the available peripherals. Although we'll go into detail on the important aspects of the chip, there are literally thousands of pages of documentation on it (the Microchip Programmer's Reference Manual alone is over 350 pages), so we are in a sense only scratching the surface. By necessity, all of what we'll look at here can be found in the standard Microchip documentation; however, the information is not always easy to find for those unfamiliar with the Microchip approach to documentation (and sometimes not even for those of us who are). The value of this chapter is not that it reveals new information (it doesn't) but rather that it organizes the relevant data in an easily assimilated format. With this knowledge, we will be in a position to use the chip wisely and will have a solid base from which to delve more deeply into the chip's features when and as we need to do so.

In the discussions that follow, we will occasionally get a cold dash of reality in the form of deviations from what the dsPIC DSC is supposed to do and what the silicon as implemented actually does. These deviations, known as *errata*[2], can drive a developer to levels of frustration usually reserved for the damned, because they cause the chip to behave in ways that differ from the documentation. From a designer's perspective there are two types of errata: those with work-arounds (that's the technical term) and those without. Errata that have work-arounds (coding or hardware design techniques that alleviate the problem) are inconvenient but not fatal; those that don't have work-arounds, particularly in an area required by an application, can cause the project to be significantly delayed, to be more expensive, or in the worst case, to fail.

That being the case, the smart designer always checks for any errata that are posted for the specific chip that is being used or for the family as a whole. Individual semiconductor companies handle the distribution of errata information differently, but Microchip is generally pretty good about assembling all known problems in errata sheets that accompany the corresponding part's data sheet or programming reference guide. These errata are posted on their website at *www.microchip.com* along with the data sheets. By checking for any errata before performing a design and during the debugging phase, the developer can save herself, her company, and her customers much frustration, delay, and expense. While it's always a good idea to assume, at least initially, that any problems lie in the developer's code or hardware design, it's also wise to check the errata sheet or with the factory if one has worked diligently on a problem and it appears that the chip simply isn't acting correctly. You might just be correct!

3.1 The dsPIC DSC's Data Processing Architecture

The dsPIC family of products is designed specifically for the high-speed processing of mathematically intensive operations, with dedicated hardware elements that handle time-intensive processes with minimal loading of the core processor. But just as the components of a mechanical engine must work together in a prescribed manner to generate power, so too must the operation of the individual elements of the dsPIC DSC be coordinated, or the chip's performance will fall far short of its capabilities. To create the necessary harmony, though, the designer must thoroughly understand the architecture of the chip and the various elements available in it.

The dsPIC DSC Memory

An understanding of the dsPIC DSC begins with its memory architecture, since it is one of those elements specifically designed to address the data-throughput requirements of a digital signal processing system. The dsPIC DSC employs a modified Harvard architecture that has separate program memory and data memory busses, allowing the processor to simultaneously fetch both an instruction and the data upon which that instruction will operate.

In a pure Harvard architecture, these two busses are completely separate, with no way to pass data between them. Although an entirely logical approach, the pure Harvard architecture is inadequate for the needs of many digital signal processing applications because it means that only a single data value can be retrieved in one cycle. Frequently, what's really required are two data values, the sampled data and a coefficient by which it will be multiplied, and in the pure Harvard architecture this requires two separate instruction cycles. To surmount this constraint, the modified Harvard architecture actually supports three busses: one to fetch the instructions (the program memory bus) and two to fetch associated data values (often referred to in the literature as the X- and the Y-memory busses). This allows true single-cycle operation, effectively doubling system throughput when compared to a pure Harvard architecture running at the same speed. Figures 3.1a and 3.1b illustrate the difference between a pure Harvard memory architecture and a modified Harvard architecture.

As we mentioned in the chapter's introduction, digital signal controllers marry some of the mathematical processing power of a true DSP with the deterministic behavior of a microcontroller, and this marriage is reflected in the dsPIC DSC's use of slightly different memory models when executing mathematically intensive

Pure Harvard Architecture

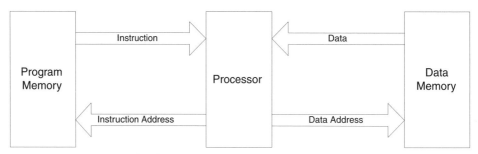

Figure 3.1a. Pure Harvard Architecture

Modified Harvard Architecture

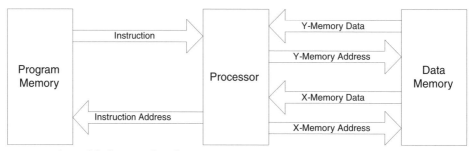

Figure 3.1b. Modified Harvard Architecture

instructions (known as *multiply-accumulate* or *MAC class instructions*) and when executing all other types of instructions (sometimes referred to as microcontroller class instructions). By doing so, the dsPIC DSC is able to employ the memory structure best suited to a particular type of instruction; those that require multiple data sources for the mathematical engine gain the performance advantage of having two data busses, while the instructions that don't have that need can view the processor as having a single unified data space. The discussion at the end of Section 3.1 on *Addressing Modes and the Address Generation Units* describes the hardware components used to access memory using the two structures.

The dsPIC DSC does lack one aspect commonly found in pure DSPs and other high-performance microprocessors: a multistage instruction pipeline. The purpose of an instruction pipeline is to maintain an internal queue of commands to the processor that can be pre-decoded to speed execution. While this approach can be and is applied successfully in a number of different applications in which the data is processed continually as a constant stream without interruption, pipelining can actually hurt processor throughput when the system must change its instruction-processing flow frequently (as occurs with interrupt-driven systems). When the processing sequence changes significantly (for instance, when the processor has to branch to a location outside of the pipeline), the pipeline has to be emptied of its existing contents and loaded with the new, correct instructions before it can continue execution. Resorting to a transportation analogy again, it's similar to driving a car on the highway; as long as one stays on the highway and goes the speed limit, one can cover a lot of ground quickly. If the driver has to constantly exit and reenter the highway, it takes much longer to travel the same distance.

Although the dsPIC DSC lacks a multistage instruction pipeline, it does utilize a single-stage instruction prefetch mechanism that reads and partially decodes instructions one cycle before they are to be executed. This allows most instructions to execute in a single cycle while significantly enhancing the deterministic timing characteristics of the system, because it is much faster to reload should the anticipated instruction not be the one to execute.

Data Space Memory Map

Data memory in the dsPIC device consists of three broad classes of memory: Special Function Registers, static RAM (SRAM), and program memory that is mapped to the data memory space.

Special Function Registers

The dsPIC DSC employs memory-mapped registers to configure, control, and monitor various aspects of the device's operation. Known as Special Function Registers, or SFRs, these registers reside in the lower 2 KB of the data space memory map. Unlike standard data space memory locations that are used for general storage, the data in the SFRs are bit-mapped, so that writing to or even reading from certain locations will cause the DSC to take a particular action.

For example, as we'll see later, transmitting a byte of data through the chip's serial port requires that the application configure the serial port's operating parameters by writing the appropriate data to the serial port's SFRs in a specific sequence. Similarly, the application can check whether new data has been received by querying the serial port's Status Register, one of the peripheral's SFRs.

The key point of dealing with SFRs is that one must be very careful when accessing the memory location because reading from or writing to any of the SFRs will affect the operation of the dsPIC device hardware and may have unintended consequences if performed improperly.

Static RAM (SRAM)

In the dsPIC30F6014A, the static RAM section extends from 0x0800 to 0x27FF, for a total of 8 KB of random access memory. Unlike SFRs, reading and writing to this memory section does not affect any aspect of the dsPIC chip's operation other than the contents of the data stored in a particular location. Applications use the SRAM section to store data that will change during the execution of the application, for instance, variables used for filtering or buffers that hold received communication data.

Under certain conditions that we'll discuss later, the SRAM can be split into two independent sections, an X-data section and a Y-data section, for improved data throughput. The starting and ending addresses of these two sections are fixed in hardware for each dsPIC device, with the X-data section encompassing the memory from 0x0800 to 0x17FF and the Y-data section running from 0x1800 to 0x27FF in the dsPIC30F6014A chip.

Program Space Memory Mapped as Data Space Memory

In some algorithms, particularly those used for digital signal processing, it's often necessary to multiply fixed coefficients by variable data. Since the coefficient data is not changing, it would save precious SRAM space if the fixed coefficient could be stored in program memory but still be accessible to the data processing hardware. Conveniently, the dsPIC architecture supports this feature, which will be described in greater detail in the Program Memory Space section.

Program Space Memory Map

Because the memory configurations for devices in the dsPIC family vary according to the specific resources available on the individual device, the designer has to refer to the corresponding data sheet for the chip's exact memory map. The memory map described here applies to the dsPIC30F6014, although most aspects of the map apply to the entire family with minor differences in specific memory address locations.

As noted previously, the dsPIC DSC supports both a program address space and a data address space, with the data space being split into an X- and a Y-data space for certain instructions. Program space memory consists of up to 4M of 24-bit instruction words, though not all of this address space is available to the user. The program memory itself is divided into the User Memory space (000000H–7FFFFEH) and the Configuration Memory space (800000H–FFFFFEH) as shown in Figure 3.2.

In operation, the application may access only the User Memory space except when using the Table Read (TBLRD) and Table Write (TBLWT) instructions to read and write the Device ID, the User ID, and certain device configuration bits found in the Configuration Memory. The Configuration Memory space is primarily reserved for use by Microchip for testing purposes and to store certain useful device identification information. To maintain compatibility with the data space addressing, which will be discussed shortly, the program space addresses are incremented by two between successive program words.

Program Memory Space Map
For dsPIC 30F6012A/6014A

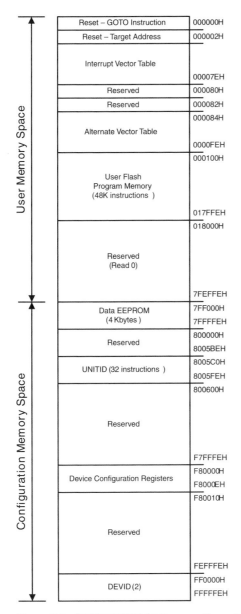

Figure 3.2. dsPIC 30F6014A Program Space Memory Map

The base of memory (address 0) holds a GOTO instruction followed by the address of the application's startup code entry point (at address 2). This two-word sequence is executed whenever the processor is reset, and it basically tells the processor to jump to the beginning of the startup code to launch the application. The Interrupt Vector Table containing the addresses of the routines that service the interrupt conditions comes next and resides in the space from address 4 to address 7EH. Following two reserved instruction words at 80H and 82H, the memory map continues with the Alternate Vector Table from 84H–FEH, the use of which is described in Section 3.2, *Interrupt Structure*.

The block of program memory from 100H–17FFEH holds the User Flash Program Memory, which supports up to 48K instructions (144 KB). This is followed by a large section of reserved memory (18000H–7FEFFEH) which is read as all 0s, and the Data EEPROM occupies the final 4K words at the top of the User Memory Space.

Accessing Configuration Memory from the User Memory Space

Normally, the application should work exclusively with memory in the User Memory space; however, as mentioned above, the application program can access certain areas of the Configuration Memory space in order to retrieve the Unit ID (32 words), the Device ID (2 words), and to retrieve or set the device configuration registers (16 words). To do this, the application must set bit 7 of the Table Page (TBLPAG) SFR to 1 and then either read or write the desired address using a Table Read (TBLRD) instruction (to read the data from the Configuration Memory) or a Table Write (TBLWT) instruction (to write the data to the Configuration Memory).

Mapping Program Memory to the Data Space

Frequently, DSP applications will store constant data in the nonvolatile program space memory to free up precious RAM locations in the normal data space. The dsPIC DSC's modified Harvard architecture supports the ability to use program memory in this way either by using the TBLRD/TBLWT instructions mentioned earlier or by mapping a 16 Kword (32 KB) program space page into the upper half of the data space using the Program Space Visibility Page (PSVPAG) register. Because the dsPIC DSC's architecture employs 24-bit (3-byte) instruction words but 16-bit (2-byte) data words, the application must account for the difference in byte alignment, and the methods employed differ based on whether TBLRD/TBLWT instructions are employed or the Program Space Visibility Page register is used to remap the memory. Although the description of the mechanics of how the application uses the two approaches is deferred to *Addressing Modes and the Address Generation Units* at the end of Section 3.1, it's important to understand the conditions under which each is best employed.

If the data being accessed must be both read and written, the application must use the Table Read/Write method; Program Space Visibility (PSV) permits only data reads. However, PSV data access is faster than the Table Read/Write method, and it is ideally suited for situations in which a constant value is being used to scale a dynamic value, as is the case when implementing a digital filter or performing a fast Fourier transform (FFT). Another limitation of PSV access is that it can be read only from the X-data space, but in practice this is not much of a constraint, since we can place the dynamic data in the Y-data space for use with the MAC class of instructions described in the next section. The Table Read/Write method is particularly useful, indeed required, when accessing locations in the Configuration Memory Space.

The DSP Engine

The dsPIC DSC incorporates a powerful DSP engine for performing the multiply-accumulate (MAC) operations that are the foundation of many signal-processing algorithms. Only certain instructions can make use of the DSP engine, and those execute with restrictions on the sources of the data they process, but the restrictions are minimal and permit the engine to read two operands, perform a MAC operation on them, and then store them back to memory, usually in a single cycle. To obtain this type of throughput, the dsPIC's designers incorporated dedicated hardware, including:

- a 17-bit × 17-bit fractional/integer multiplier,

- two 40-bit accumulators,

- a 40-stage barrel shifter that can shift up to 16 bits left or right in single instruction, and

- dual address generation units (AGUs) that can calculate modulo addressing (both AGUs) and bit-reversed addressing (one AGU).

Properly employed, these dedicated hardware subsystems significantly reduce the processing required to implement complex signal-processing algorithms. A block diagram of the DSP engine components is shown in Figure 3.3. Before discussing the hardware components, however, we need to examine the way in which numerical data is represented in the dsPIC DSC.

DSP Engine Block Diagram

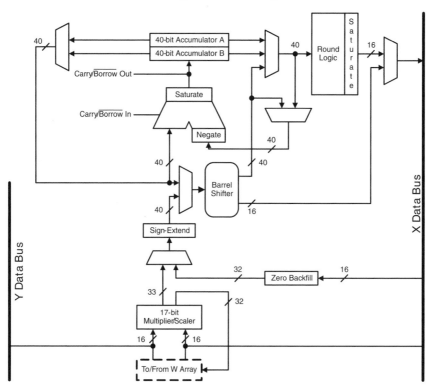

Figure 3.3. dsPIC30F DSP Engine Block Diagram

Numeric Data Representation

The dsPIC DSC can operate on signed data that uses either a fractional or integer representation. Signed integer data uses the standard two's complement format, in which the most significant bit (MSB) is the sign bit and each subsequent bit is a power of two, as shown in Figure 3.4.

-2^{15}	2^{14}	2^{13}	2^{12}	2^{11}	2^{10}	2^9	2^8	2^7	2^6	2^5	2^4	2^3	2^2	2^1	2^0

Figure 3.4. 16-bit Signed Two's Complement Integer Representation

An N-bit two's complement integer can represent values between -2^{N-1} and $2^{N-1} - 1$. A 16-bit two's complement integer number can take on any value between $-32,768$ (0x8000) to $32,767$ (0x7FFF), while a 32-bit number can range from $-2,147,483,648$ (0x80000000) to $2,147,43,647$ (0x7FFFFFFF).

When working with two's complement integer data, it's important to make sure that the sign bit is properly handled when performing addition or subtraction operations on numbers with different bit widths, or computation errors may occur. For example, if we were to add the 8-bit two's complement value of –4 (0xFC) to the 16-bit value of 20 (0x0014) without sign-extending the 8-bit value, we'd get a result of 0x00FC + 0x0014 = 0x0110 or 272 decimal, a far cry from the 16 (0x0010) we expected. If, however, we sign-extend[3] the 8-bit value before the addition, we get a result of 0xFFFC + 0x0014 = 0x10010 or 0x0010 if we maintain our 16-bit representation length (which would cut off the MSB of the result). The rule for adding two different-sized two's complement values is to sign-extend the length of the shorter representation to match the length of the longer representation. Note that what we're interested in is the length of the representation, not the length of the specific value we're using. For instance, if we are adding an 8-bit value to a 16-bit value, we need to sign-extend the 8-bit value to sixteen bits, regardless of the actual values contained in either the 8-bit or the 16-bit representations.

Although integer data is very fast, it suffers from the tendency for multiplications to overflow fairly easily, which normally requires that the application include additional software to check for and to handle the overflow to prevent erroneous results. Multiplying an M-bit integer by an N-bit integer yields an (N + M)-bit value. It doesn't take too many multiplications to overflow even a large multiplier. One way to avoid the overflow problem is to use fractional arithmetic, in which all values are constrained to the range –1 to +1. In this case, multiplications are guaranteed to stay within the same range, eliminating the problem of overflow (at least for the multiplication portion of a mathematical operation). Because this approach is quite common, the dsPIC DSC can be configured to perform signed fractional arithmetic operations using the QN format. In the QN format, the data is represented as a two's complement fraction in which the MSB is defined as the sign bit and the radix point (the decimal point when using a base-ten system) is implied to be immediately after the sign bit. The "N" in the QN notation simply specifies the total number of bits in the number. A Q16 number uses 16 bits to represent values, while a Q32 number uses 32 bits. The literature also refers to this format as 1.(N–1) format, with the "(N–1)" term representing the number of bits used to represent the mantissa (the fractional portion of the number). Using this notation, a Q16 number would be referred to as having a 1.15 representation, and a Q32 number would use a 1.31 representation. Figure 3.5 shows an example of a Q16 or 1.15 fractional representation.

-2^0	2^{-1}	2^{-2}	2^{-3}	2^{-4}	2^{-5}	2^{-6}	2^{-7}	2^{-8}	2^{-9}	2^{-10}	2^{-11}	2^{-12}	2^{-13}	2^{-14}	2^{-15}

Figure 3.5. Signed Q16 or 1.15 Fractional Representation

A signed QN value can represent numbers in the range -1 to $1 - 2^{1-N}$. Q16 numbers, then, can represent values between -1 and 0.999969482421875, while Q32 numbers can represent values between -1 and $(1 - 4.66 \times 10^{-10})$, essentially -1 and 1 for all practical purposes.

As with two's complement integer values, addition or subtraction using QN fractional numbers requires that the data be properly extended before performing the operation. In this case, however, the extension is handled by zero-padding to the right of the LSB in the mantissa of the smaller representation rather than by sign extending, as is the case for integer arithmetic. For example, to add a Q8 number to a Q16 number, one first pads the right of the Q8 value with 8 additional zero bits to form the corresponding Q16 value.

Multiplication of an M.N fraction by an R.S fraction produces an (M + R).(N + S) result. When two 1.15 values are multiplied together, then, the result is a 2.30 number. This is analogous to the situation for multiplying a decimal fraction with N digits to the right of the decimal point by a second decimal fraction with S digits to the right of the decimal point, with the result having N + S digits to the right of the decimal point. In fractional multiplication, however, the multiplicands are sign-extended prior to multiplication (i.e., negative values have a "1" prepended to the most significant bit, positive values have a "0" prepended), and the result is left-shifted by a single bit position to correctly handle the sign and fractional bit alignments. An example will illustrate this more concretely.

Example 3.1:
To correctly multiply –0.25 by 0.25 using Q8 (1.7) notation, the steps are:

1. Sign-extend the two values to be multiplied:

    −0.25:   1.110 0000b → 1 1.110 0000b

    0.25:   0.010 0000b → 0 0.010 0000b

2. Multiply the two values together:

    1 1.110 0000b

    × 0 0.010 0000b

    11 1100 0000 0000b

    000 0000 0000 0000b

    0000 0000 0000 0000b

    0 0000 0000 0000 0000b

    0 0001 1110 0000 0000b

3. Left-shift the result by 1 position:

0011 1100 0000 0000b

4. Place the implied decimal point prior to the 13th least significant bit:

001.1 1100 0000 0000b = −0.0625 decimal

Hardware Multiplier

The 17-bit × 17-bit hardware multiplier allows the dsPIC DSC to multiply two signed 16-bit values together using either fractional or integer arithmetic. It should be noted that, with the exception of multiplication, the operations for integer and fractional data are identical; the only difference is how they are interpreted. For example, if two 16-bit two's complement integer values, say 0x2001 and 0x17F0 are added together, the result will be 0x37F1. If the same two signed Q16 values (0x2001 and 0x17F0) are added together, the result is also 0x37F1. The difference is that 0x37F1 as a 16-bit two's complement integer represents 14,321 decimal, while that same value as a signed Q16 number represents approximately 0.437042236.

As noted above, multiplication of signed fractional two's complement numbers requires the sign-extension of those values by one bit, hence the need for a 17-bit × 17-bit hardware multiplier for multiplying two Q16 values. To handle the special case where two Q16 inputs with a value of 0x8000 (−1) are multiplied together, the hardware multiplier automatically corrects the resulting value to be 0x7FFFFFFF, the closest value to +1. If the hardware did not handle the error in this way, the resulting overflow would produce the incorrect value of −1 × −1 = 0. Although the correction introduces a small error, that error is much less significant than allowing the multiplier to overflow.

Because the DSP engine handles integer and fractional multiplication differently, the application must specify which numerical notation is being employed. It does so by configuring the IF bit in the CORCON (Core Configuration) SFR; when the flag is cleared (set to "0"), multiplications are handled using fractional notation, and when the flag is set (equal to "1"), the multiplier performs standard integer multiplication. When in fractional mode, the multiplier automatically includes sign-extension and the requisite left-shift to maintain proper radix point alignment. If the application performs either a MAC or a MPY instruction (the two DSP multiply instructions) using fractional data without first clearing the IF flag, the application must explicitly perform the left shift or the resulting value will be incorrect.

The same multiplier is used for the standard MCU multiply instructions (variations of the MUL command), which are either 8-bit or 16-bit integer operations (both signed and unsigned). When both operands are 8 bits, the result is a 16-bit value, while 16-bit operands produce a 32-bit result.

Dual 40-bit Accumulators

The output of the multiplier is routed to the data accumulator, a subsystem consisting of a 40-bit adder/subtractor with sign extension logic and two 40-bit accumulators (creatively designated as "A" and "B") that can serve as both source and destination for the accumulation operation. For certain instructions, the ADD (add accumulators) and LAC (load accumulators) commands, the data can also be scaled by the barrel shifter before performing the accumulation operation, which can be a convenient way to normalize the data. The provision for two accumulators is particularly useful when performing complex number arithmetic, a common DSP requirement.

One might reasonably ask why the data accumulator is 40 bits wide when the output of the multiplier is a 32-bit value. The answer is that the wider accumulator provides some additional computational "room" to avoid problems with overflowing the accumulator when performing multiple accumulations, such as might be encountered with a multitap filter. Recall that the reason we often use fractional arithmetic is to avoid problems with overflow when multiplying; by limiting the range of input operand values to be between –1 and +1, we're able to ensure that any output result is also between -1 and +1. Unfortunately, we have no corresponding way to limit the output value of an addition or subtraction operation; summing two fractional values together always has the potential to create a result that is outside our –1 to +1 range. The only way to prevent overflow is to provide "guard" bits to the left of the radix point, with the more guard bits the better. In the case of the dsPIC DSC, the chip's designers have provided 8 guard bits (the difference between the size of the 40-bit accumulator and the 32-bit inputs), which allows up to 256 consecutive additions of full-scale values at either end of the input range (i.e., either adding two –1 values or two +1 values) before the data accumulator will overflow. In practice, depending on the values actually being accumulated, these 8 guard bits may allow far more operations before overflowing, or the accumulator may never overflow at all.

Because arithmetic overflow can be a huge problem, the dsPIC DSC offers two flags per accumulator to allow the application to check for this condition. These flags, found in the CORCON register, indicate whether either of the two following conditions has occurred:

1. the accumulator has overflowed beyond bit 39 (indicated by the SA or SB flag being set), and/or

2. the accumulator has overflowed into the guard bits (indicated by the OA or OB flag being set).

The dsPIC30F Family Reference manual correctly describes the first of these conditions as "catastrophic," because when it occurs, the sign bit for the accumulator is destroyed. The second condition is not too serious, but it does indicate that there is potential for subsequent accumulations to generate a catastrophic overflow.

Catastrophic overflow can cause very serious problems when it occurs, because outputs from such an operation will wrap either from positive to negative or from negative to positive. This represents a mathematical discontinuity, which violates the first tenet of most digital signal processing (and certainly the DSP we're discussing in this book), namely that the systems can be modeled mathematically as linear systems. In most microprocessors, checking for and correcting catastrophic overflow is difficult, time-consuming, and code-space expensive, but the dsPIC's designers included special configurable logic that not only detects catastrophic overflows but allows the data accumulator to handle them gracefully without any additional software overhead. When configured to do so, the dsPIC DSC will convert what would normally be a catastrophic overflow into either the most negative or most positive value (depending on the direction of the overflow), a situation known as *accumulator saturation*. In this respect, the dsPIC DSC mimics how an analog multiplier would handle overflow, saturating the output of the multiplier rather than wrapping around. The dsPIC DSC's saturation behavior can be tailored to saturate upon overflow of either its natural (32-bit) or extended (40-bit) range, depending upon the application.

The dsPIC DSC offers considerable flexibility in the handling of the overflow detection flags and processes the two types of flags somewhat differently. Both types of flags (OA/OB and SA/SB) are modified each time that data passes through the adder/subtractor, but the SA/SB flags can only be cleared by the application (the OA/OB flags are automatically cleared by hardware when the overflow condition clears). In addition, the setting of either type of flag can optionally generate an Arithmetic Warning trap when the corresponding overflow trap enable bit (OVATEN/OVBTEN for guardbit overflow or COVTE for catastrophic overflow) is set in the Interrupt Control 1 register (INTCON1). This allows the user to handle such error conditions immediately, for example by reducing system gain to eliminate the problem.

To further reduce the software overhead required to detect and process overflow conditions, the dsPIC DSC also has two flags that are the logical OR of either the OA and OB bits (the OAB flag) or the SA and SB bits (the SAB flag). These two flags are found in the Status register (SR) and allow the application to quickly check for overflow in either accumulator, which is helpful when performing complex arithmetic computations.

40-bit Barrel Shifter

The 40-bit barrel shifter provides a quick (single-cycle) method to right-shift data up to 15 bits to the right or 16 bits to the left. This feature is particularly helpful in normalizing data to a specific range or when aligning a serial bit stream from one of the communication ports. Data is fed into the barrel shifter from the X bus between bits 16 to 31 for right-shift operations and between bits 0 to 15 for left-shift operations.

Addressing Modes and the Address Generation Units

One of the key reasons for the dsPIC DSC's high data throughput is its ability to employ a number of flexible addressing schemes to access and process data efficiently. Not only does the chip offer numerous addressing modes, it also employs hardware-based address generation units (AGUs) that can implement overhead-free modulo and bit-reversed addressing that drastically reduces the software overhead associated with complex signal-processing algorithms. Prudent use of the AGUs' features and the various addressing modes can significantly trim the code space and execution time required to implement such algorithms.

The dsPIC DSC supports four basic data addressing modes, some of which may be extended by certain instructions:

1. File Register or Memory Direct

2. Register Direct

3. Register Indirect

4. Immediate Operand

The File Register or Memory Direct addressing mode embeds the 13-bit absolute memory address within the instruction. Since this mode only permits 13 address bits, it can access only the first 8 KB of the data space, which is also known as the "near" data space. Probably the most common usage of this addressing mode is to access the Special Function Registers (SFRs) located in the lower 2 KB of the data space. Note that near addressing cannot be used to access the upper 2 KB of the SRAM section.

Register Direct addressing allows direct access to all of the registers in the W register array, and the instructions that support this mode specify the operand source or result destination register as a particular W register (e.g., W1, W4, etc.). This mode is particularly useful for three-operand instructions of the form $Z = X + Y$, where X, Y, and Z are all W registers and the operation could be addition, subtraction, or a bit operation. Note that in the case of three-operand instructions, the two source operands must be different registers, although the destination register may be any of the registers.

By implementing the Register Indirect addressing mode, the dsPIC designers made it very easy to support higher-level languages, particularly C, that employ pointers to data. The hallmark of Register Indirect addressing is that the address of the data to be accessed is stored in one of the 16-bit W registers. This offers two significant advantages over the other addressing modes: it permits the application to access a larger data space (64 KB instead of the 8 KB of File Register or the 32 bytes of Register Direct), and it allows the application to modify at run-time the address of the data to be accessed.

Of the basic addressing modes, Register Indirect is the most flexible, allowing the application to increment or decrement the contents of the source operand register and/or the destination result register as part of the single-cycle instruction execution. The increment/decrement operation can be performed either prior to retrieving the addressed data or after retrieving it, which allows the use of a wide variety of powerful looping constructs. In addition, some instructions also allow the address register contents to include either a register offset (i.e., combine the address register contents with the contents of another register to obtain the target address) or a literal offset (i.e., combine the address register contents with a fixed value to obtain the target address). These last two variations allow an application to efficiently access data from fixed tables or from structures.

Finally, Immediate Operand mode embeds the data value in the instruction itself, with the size of the data that can be used dependent upon the specific instruction.

3.2   Interrupt Structure

Interrupts are asynchronous changes in the flow of instruction execution that are caused either by an external event (e.g., a hardware signal changing state) or by an internal condition (the expiration of an internal timer, for example). Technically, the dsPIC DSC differentiates between true interrupts, which are generated by an anticipated event whose processing is a normal part of the application, and traps (or processor exceptions) that handle erroneous processing conditions that should not occur in the course of normal operation. In practice, both interrupts and traps are processed in the same manner, so this discussion will usually lump the two under the single banner of "interrupts."

The dsPIC DSC supports a total of 45 interrupt sources, four types of traps, and six conditions that can reset the processor. Because most applications require the use of only a few of these exception handlers, the dsPIC DSC's interrupts must be enabled individually, with a Global Interrupt Enable flag that allows the program to quickly turn off all interrupts with a single operation. When an interrupt event occurs whose enable flag is set and as well, the Global Interrupt Enable flag is set, the processor will store the program address from which it is currently executing on the interrupt stack and retrieve the address of the appropriate interrupt handler function (known as the *interrupt vector*) from the corresponding entry in the Interrupt Vector Table (IVT) located between address 000004H and address 00007FH of the program space. Upon retrieving the interrupt handler address, the dsPIC DSC jumps to that address and begins executing code until it encounters a Return from Interrupt Enable (RTFIE) instruction, at which point it will retrieve the address most recently stored on the interrupt stack and resume normal processing from that point.

For simplicity, the preceding discussion leaves out a crucial aspect of the dsPIC DSC's handling of interrupts, namely that the interrupts are *prioritized* into one of seven different levels, with level 0 having the lowest priority and level 6 the highest. This means that the interrupts have a built-in hierarchy, with some interrupt sources being deemed more important that others. Interrupt sources whose vectors are located at lower addresses in the IVT are considered to have higher priority, and the priority order in which the vectors appear is considered to be the vectors' "natural" priority. Because the dsPIC's designers realized that not all applications will require the same priority for different interrupts, the priority of individual interrupts can be changed by the application code at run-time to be either higher or lower.

One of the key aspects of interrupt processing is the amount of time it takes from the point at which the interrupt condition becomes active and when that condition is actually acted upon by the processor, a value known as the *interrupt latency*. The goal with interrupt latency is two-fold: to minimize the time it takes to start processing the interrupt and to keep the latency period as steady as possible so that the processing of individual components in the application can be handled reliably. In the dsPIC DSC, these twin goals are met superbly, with a fixed latency of five instruction cycles from the point at which the interrupt request occurs (i.e., the time when the interrupt source becomes active) and the point where the interrupt service routine (ISR) actually starts. It's important to note, however, that this latency assumes that no other interrupts are being processed at the current priority level or higher; if there are others being serviced at the same or a higher priority level, the processing for those ISRs will complete before the new interrupt is serviced and hence the latency can increase well beyond the five-cycle base.

Shadow Registers

The earlier discussion also left out an important performance-enhancement feature of the dsPIC DSC: the use of shadow registers to provide quick context switching when interrupts or other forms of exceptions occur. When an interrupt occurs, the application can store the contents of certain flags in the Status Register as well as the contents of the TBLPAG, PSVPAG, and W0–W14 registers to the shadow registers with a call to a single instruction, PUSH.S. Upon completion of an interrupt that uses the shadow registers, the ISR must call POP.S to restore the standard registers to the contents stored in the shadow registers.

Because the shadow registers are only one level deep, it is critically important that the application never attempts to make two consecutive calls to PUSH.S or POP.S without an intervening call to its complement or the original register contents will be lost. This means that ISRs for lower priority interrupts cannot use the shadow registers if ISRs for higher priority interrupts also use those registers, since the occurrence of one of the higher priority interrupts during processing of a lower priority interrupt would result in the shadow register contents being overwritten.

3.3　The On-chip Peripherals

One of the dsPIC DSC's greatest strengths is the level of system integration that it brings to sensor applications. Depending on the particular version, a single dsPIC DSC can incorporate on-board peripheral modules that directly digitize analog input signals or read digitized data from external A/D converters, that implement multiple free-running and/or event-based timers, and that support several industry-standard communication protocols. Since the dsPIC's architecture allows the designer to specify the processing priority of the individual peripherals at run time, they are able to optimize the chip's operation and throughput for the specific application.

In the following discussions of the various peripheral modules, the intent is to clarify the information found in the dsPIC data sheets and to establish checklists that the designer can employ to ensure that each module is used properly. The discussions also incorporate insights available from other Microchip sources such as tutorials and application notes, with the goal of creating a coherent body of knowledge out of the wealth (one might easily say glut) of raw information about the chip.

Data Acquisition Peripherals

In any intelligent sensor system, the "digital" part of the DSP signal chain[4] starts when the input analog signal is digitized by a device called, imaginatively, an analog-to-digital converter (also known as an *A/D* or an *ADC*). The dsPIC DSC supports a number of different ways to perform signal digitization:

- through an on-board multichannel ADC,

- by using an external audio coder/decoder (codec) whose output is then read through the dsPIC's Data Converter Interface (DCI),

- via an external ADC controlled through either the Serial Peripheral Interface (SPI) port or the Inter-Integrated Circuit (I^2C) ports.

In this book, we will focus on the first method, the use of the on-board ADC.

Analog-to-Digital Converters

Probably the most common and certainly the most cost-effective method of digitizing the input signal is the use of the dsPIC DSC's internal ADC. As of this writing, all dsPIC30F series devices support between 6 and 16 channels of ADC inputs with either 10-bit or 12-bit resolution, depending upon the particular chip. Devices that offer 10-bit resolution are generally targeted for motor control or video-processing applications that can tolerate the reduced resolution (only 1024 unique levels vs. the 4096 levels provided by a 12-bit ADC), while those with 12-bit resolution are intended to service the general sensor market. The dsPIC30F6014A, the device that we'll use in the applications we develop throughout this book, supports 16 channels of 12-bit conversion.

To get a feel for what sort of ramifications these different resolutions might have in real-world applications, consider the two separate cases of monitoring the rotational position of a steering wheel in a steer-by-wire system and of measuring the pressure inside the mold of an injection molding machine. In the case of the steering wheel, let's assume that we're able to determine the rotational position of the wheel using the full resolution of the ADC to measure 360° of rotation. Under those conditions, a 10-bit ADC would allow us to measure the steering wheel's position to within approximately one-third of a degree (360° / 2^{10} levels = 360° / 1024 = 0.35°/level), while a 12-bit ADC would be able to resolve the rotational position to within less than a tenth of a degree. Either of these would be perfectly adequate for most steer-by-wire systems, being well within the limits of human perception (assuming, of course, that we're using a human driver).

The situation is not nearly so clear-cut when monitoring a large-ranged signal such as pressure inside a molding machine, which usually varies between 0 and perhaps 30,000 psi.[5] In this case, each level in the 10-bit ADC corresponds to just under 30 psi, while each level in the 12-bit ADC corresponds to just over 7 psi. That may not seem particularly significant, but if the molder is using the pressure at a particular point in the mold to open a gate to allow more plastic into the mold, the difference in resolution can mean the difference between a stable, profitable molding process with low scrap rates and an unstable, unprofitable process with unacceptably high scrap rates.

As Richard Fischer of Microchip points out in his excellent webinar on the dsPIC's ADC module,[6] the ADC performs two major operations every time it converts an input analog signal to the corresponding digital value. First, the ADC samples the analog input and stores the *analog* sample in its sample-and-hold amplifier, a process known as *acquisition*. As Richard notes, capturing the analog sample is analogous to taking a photograph of the analog waveform, with the acquisition time being similar to the exposure time of a photograph. The application determines how much time each acquisition is allowed, but it must ensure that the acquisition time is sufficient to accurately capture the signal, just as the photographer must ensure that the shutter speed of his camera allows sufficient light to reach the film (or CCD array in this era of digital cameras) for a good picture.

Once the signal has been captured, the ADC performs the second major operation: it converts the sampled analog signal into the corresponding numeric digital value, a process generally referred to as *conversion*.[7] Figure 3.6 shows the relationship between sampling time (acquisition time), A/D conversion time, and the complete conversion cycle time. Depending upon how the module is configured,

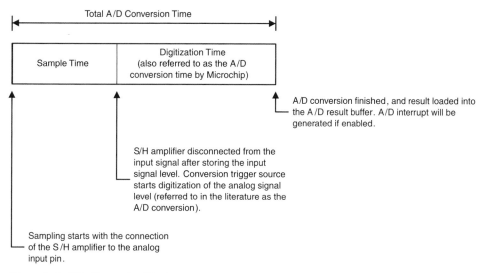

Figure 3.6. A/D Conversion Timing

the converted digital number can be in one of four different integer or fractional formats, allowing the application to use the representation that works best with its particular algorithms.

A dsPIC application can sample data using either a *polled* (or *manual*) approach in which the code explicitly initiates each sampling operation, or it can perform auto conversion sampling in which the ADC module is configured initially and then the hardware repeatedly performs the acquisition and conversion, only interrupting the processor when the specified number of channels have been converted. In most applications, interrupt-driven operation is preferred because it ensures that the data is sampled with temporal stability—i.e., the time between subsequent samples is very stable. This temporal stability is very important since most DSP algorithms assume that sampling frequency remains constant; an excess amount of jitter[8] can cause erroneous results that would not be seen if the sampling periods were more tightly controlled. Polled operation is usually used either to debug the hardware interface, when the value being measured can be "spot checked" in a nonsynchronous manner (for instance, checking the level of the oil in a car in a background "check engine" operation), or when the desired sampling frequency is too slow to be handled by the ADC module's hardware.

The dsPIC DSC uses a single successive approximation A/D multiplexed between all of its input channels to perform the digitization. Depending on the ADC's resolution and the specific device, the input channels are multiplexed through either one, two, or four sample-and-hold (S/H) amplifiers before being converted to a digital value, as shown in Figure 3.7. Only the 10-bit ADCs support multiple S/H amplifiers; the

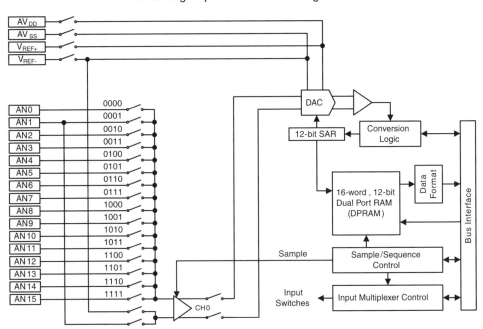

Figure 3.7. dsPIC Analog-to-Digital Conversion Circuitry

12-bit ADCs are limited to a single S/H circuit. At first blush, the ability to use multiple S/H amplifiers might not seem that important since their outputs still must be converted by a single conversion circuit, but they do offer an important feature: the ability to make truly simultaneous samples. Often, this feature is nice to have but not strictly necessary for a particular design, but in applications where it is required, the dsPIC DSC's on-board multiple S/H amplifiers are a tremendous advantage.

The internal A/D can sample a maximum of just under 200 Ksps (200,000 samples per second) across the full supply voltage, but that sampling bandwidth is spread over all of the channels, so each channel can digitize at a maximum of 12,500 sps (200,000 sps / 16 channels = 12,500 sps/channel) when all channels are used. In practice, of course, this sample rate may have to be reduced to allow sufficient processing time for any conditioning and analysis algorithms required by the application. The sampling rate is also dependent upon the supply voltage that powers the chip; the device is capable of a rate just under 200 Ksps when the chip operates with a supply voltage between 4.5V and 5.5V. If the chip is run at a lower voltage, the maximum sampling rate drops to 100 Ksps.

As Figure 3.7 shows, the ADC circuitry offers a great deal of flexibility, but this flexibility requires significant run-time configuration on the part of the application in order to function properly. The ADC module employs six 16-bit status and control registers and an additional 16-word dual port read-only buffer (ADCBUF0 through ADCBUFF) that contains the results of the conversions.

In this discussion, we will attack the configuration issue by sequentially exploring the tasks we need to accomplish in the order in which they need to be performed rather than by simply listing which bits do what in each register. This approach will hopefully give the reader a better understanding of the overall process than might be obtained with a checklist of register bit assignments, though we'll conclude with just such a checklist for the sake of completeness.

Before configuring the ADC module, we first need to determine the following information:

1. Which I/O port pins are to be analog inputs, which are to be digital inputs, and which are to be digital outputs?

2. How fast do we need to sample the analog inputs?

3. Do we need to use interleaved sampling, or can we simply generate a single interrupt after converting all of the signals during a given sampling period? Does our processing throughput require that we multiplex the results buffer?

4. In what format to we want the converted data?

Once we've answered those questions, we can proceed to the actual module configuration.

Step 1 – Signal Path Configuration

Not surprisingly, the first step in configuring the ADC module is the identification of the specific pins that are to be treated as analog inputs, digital inputs, and digital outputs. Although it's possible to change this configuration at any point during runtime, for simplicity this discussion will assume that we're configuring the I/O signal types and directions only once at the beginning of program execution. Should the application require that the signal types and/or directions be changed later in the processing, we can do so by applying the steps discussed here but only *after* interrupts have been disabled. It's an extremely bad idea to change either the type or direction of I/O signals if there's a possibility that code may be executed that assumes that the signals are configured differently.

Step 1A – Configuring the I/O Port Pins

As is standard with Microchip products, the application specifies an I/O port signal's direction through the TRISx register, where the "x" refers to the specific port (TRISA for PORTA, TRISB for PORTB, etc.). Bit assignments for the 16-bit TRISx registers correspond one-to-one with the bits in the corresponding 16-bit PORTx register. To configure a signal as an input (whether digital or analog), the application sets the corresponding TRISx bit to a "1", or to configure the signal as an output, the application clears the TRISx bit to a "0". An easy way to remember this convention is to think of the "0" as standing for "0utput" and the "1" as standing for "1nput".[9] Figure 3.8 shows the bit mapping for a generic TRISx register.

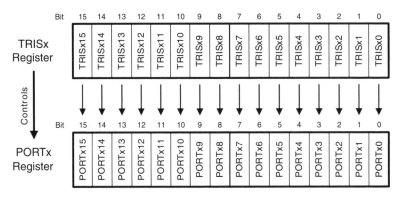

Figure 3.8. Generic TRISx Register Bit Mapping

On the dsPIC30F6014A, all of the analog inputs (denoted as AN0 to AN15 on the data sheet) are located on Port B, so for the purposes of configuring the analog inputs, we're concerned with the register TRISB. If we want to digitize the signal on a particular Port B line, we have to set the corresponding bit in TRISB to a value of "1". For example, if we want to digitize the signals on AN0, AN1, and AN8, to use Port B<4> and Port B<6> as digital inputs, and to employ the remainder as digital outputs, we would write a value of 0x0153 to the TRISB register (0x0103 (the analog inputs) ORed with 0x0050 (the digital inputs)).

Once we've set the signal direction, we then need to specify whether a particular input is an analog or a digital signal, a task we accomplish by configuring the 16-bit A/D Pin Configuration register (ADPCFG) appropriately. Each bit in the register corresponds to one of the analog inputs (AN0 to AN15); setting a bit to a "1" configures the corresponding signal to be a digital input, and clearing a bit to "0" configures the corresponding signal as an analog input. The register's bit-mapping is shown in Figure 3.9.

Continuing the previous example, we would need to set the value of the ADPCFG register to 0xFEFC, leaving the bits for AN0, AN1, and AN8 as the only cleared bits (and hence AN0, AN1, and AN8 as the only analog inputs).

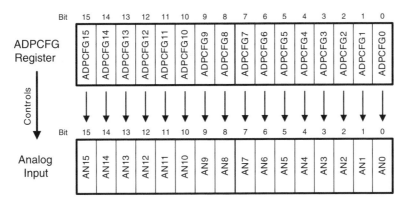

Setting an APCFG bit to '0' configures the corresponding port (AN) pin as an analog input; setting an APCFG bit to '1' configures the corresponding port (AN) pin as a digital input.

Figure 3.9. AD Pin Configuration Register Bit-mapping

What would happen if the application configured a signal as a digital output in TRISB but as an analog input in ADPCFG? In that case, the port pin would indeed operate as a digital output, but in addition we would be able to digitize the voltage level on the pin and read it in through the ADC module. That feature might be useful in a safety-critical application so that the program can determine whether a particular digital output is actually in the correct state, and not, for instance, shorting to ground or VDD.

Although this discussion will continue with the configuration of the ADC module itself, in most applications the designer would want to finish configuring the rest of the I/O port signal directions and states (i.e., those for Ports A, C, etc.). first so that the hardware would be put in a known-safe state as quickly as possible. One important aspect to note is that analog voltage levels on a pin configured as a digital input may cause the input buffer to consume more current than that for which the device is rated (that's the technical way of saying that we might destroy the chip itself).

Step 1B – Selecting the Reference Voltage Sources

With the signal directions and types configured, we move next to selecting the voltage references that we'll use for digitization. For simplicity of design, conversions are often made over the full range of the supply voltage (AV_{SS} to AV_{DD}); however, to improve the resolution of the measurements, the conversion reference voltages can be limited to the range V_{REF-} to V_{REF+}. This can significantly improve the chip's ability to measure limited-range input signals since the conversion is performed only over the voltage range of interest, not the full supply voltage range. For instance, if the designer knows that the input signal will range from 0.5 to 1.5 V, setting the lower and upper reference voltages to those two values allows the ADC to produce conversions in 0.24 mV increments (($V_{REF+} - V_{REF-}$) / 2^{12} levels = (1.5V – 0.5 V) / 4096 levels = 0.24 mV/level). If the conversion were performed using V_{REF-} of 0V and V_{REF+} of 5V, as might be the case for a standard 5V system, the resolution would be 5 times worse, or 1.2 mV / level. That might not sound so bad, but for low-voltage inputs such as that generated by a thermocouple, the difference might be huge.

It's also possible to configure the module so that one input voltage is converted relative to another input voltage, effectively producing a differential measurement. This mode is limited, however, in that the conversion is unipolar; i.e., the reference input voltage must always be less than or equal to the voltage on the nonreference input. Even with this restriction, however, the ability to operate differentially can be a useful way to reduce common-mode noise[10] on the inputs.

Recall that the upper and lower reference voltages represent the range of analog input voltages that are then mapped to the 10-bit or 12-bit ADC output values. The goal is to use a reference voltage range that most closely matches that of the input signal so that we get the most resolution in the converted digital values. With the dsPIC DSC, we have four possible combinations of upper and lower reference voltages since we can select either AV_{SS} or V_{REFL} for the lower reference voltage and either AV_{DD} or V_{REFH} for the upper reference voltage.

Setting the reference voltage sources is very straightforward; the application simply sets the Voltage Configuration bits (VCFG<2:0>) in the A/D Control Register 2 (ADCON2). Although there are three bits in VCFG and thus theoretically eight possible reference voltage configurations, setting the upper bit (VCFG<2>) to a "1" always selects an upper reference voltage of AV_{DD} and a lower reference voltage of AV_{SS}. The table in Figure 3.10 shows the reference voltage configurations that correspond to the various VCFG values, and Figure 3.11 shows the bit assignment of VCFG in the ADCON2 register.

Step 1C – Selecting the Analog Inputs to Digitize
To reduce unnecessary processing overhead, the application can specify which specific analog input channels are to be converted, the order in which they are to be converted (to a degree, at least), and the number of conversions that are to take place before interrupting the processor. This frees up a significant amount of processing time in applications in which only a subset of the available ADC input channels are needed, since the firmware doesn't have to spend time handling converted data for channels whose values will never be required. In all sampling modes, digitized data is stored in the 16-word ADCBUF buffer in the order in which it is sampled, starting with

ADC Reference Voltage Selection Using the VCFG Bits in ADCON2

VCFG<2:0>	V_{REFH}	V_{REFL}
000	AV_{DD}	AV_{SS}
001	V_{REF+}	AV_{SS}
010	AV_{DD}	V_{REF-}
011	V_{REF+}	V_{REF-}
1xx	AV_{DD}	AV_{SS}

V_{REFH} = Upper reference voltage to A/D
V_{REFL} = Lower reference voltage to A/D
V_{REF+} = External upper reference voltage
V_{REF-} = External lower reference voltage
AV_{DD} = Analog positive power rail
AV_{SS} = Analog negative power rail (ground)

Figure 3.10. ADC Reference Voltage Configuration Values

VCFG Bit Mapping In ADCON 2

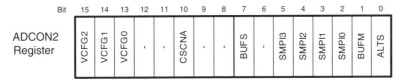

Figure 3.11. VCFG Bit-mapping in the ADCON2 Register

location ADCBUF0 and continuing until the number of samples specified in the Samples Per Interrupt (SMPI) value in ADCON2 have been digitized, at which point the next sample is stored again at ADCBUF0. Because of this behavior, the application's ADC interrupt handler must ensure that it can unload and process all of the samples it needs from the ADCBUF buffer in a single sampling time, since failure to meet this timing requirement will mean that the sample stored at ADCBUF0 will be lost (overwritten by the next sample stored there).

The ADC module can convert signals in one of two user-configurable channel patterns, and it can even combine the two approaches for more flexibility. The first technique, known as *interleaved* or *alternate channel sampling*, digitizes a specified input channel (this is imaginatively denoted in the documentation as the Mux A input), stores the digitized A value in ADCBUF0, then digitizes a second, usually different, input channel (denoted as the Mux B input) and stores it in ADCBUF1.

Sampling continues in an alternating fashion (first the Mux A input, then the Mux B input) until SMPI conversions have been stored in ADCBUF, at which point the ADC module generates an interrupt (assuming its interrupt is enabled). The next sample collected after the interrupt is stored in ADCBUF0, and the process repeats.

To enable basic interleaved sampling (i.e., alternating between two inputs), the application must set the Alternate Sampling (ALTS) bit in ADCON2 to a value of "1". It must then designate the specific channels to sample as the Mux A and the Mux B inputs by setting the 4-bit Channel 0 Select A (CH0SA) field and the Channel 0 Select B field of the A/D Channel Select Register (ADCHS) to the 0-based index of the desired Mux A and Mux B inputs, respectively. In the same write to ADCHS, the application should also select the negative input for the S/H circuit for the Mux A and the Mux B inputs by writing to the Channel 0 Negative Input Mux A flag (CH0NA) and the Channel 0 Negative Input Mux B flag (CH0NB). Clearing the CH0NA or CH0NB flag selects V_{REF-} as the negative input for the corresponding Mux input, while setting the flag selects the voltage on the AN1 input as the negative input. In this way, the ADC module can support unipolar differential mode, with the important requirement that the voltage on AN1 always be less than or equal to the selected Mux A or Mux B voltage.

The second technique, known as *channel scanning*, involves sampling data from a group of semisequential channels with the module generating an interrupt after the appropriate number of samples has been converted. We use the term *semisequential* to indicate that while the channels are converted in a fixed sequential order (from lowest-indexed channel to highest), individual channels may be eliminated from the sampling sequence if their value is not needed. As with interleaved sampling, digitized values are stored in ADCBUF as they are converted, beginning with ADCBUF0. After SMPI samples have been stored to ADCBUF, the processor generates an interrupt so that the data can be transferred from the buffer to the application. New samples following the interrupt are stored in ADCBUF, starting at ADCBUF0 even if ADCBUF was not completely filled originally.

Finally, the dsPIC DSC supports a combination of interleaved sampling and channel scanning, with the constraint that the channel scanning can only be performed during the Mux A portion of the sampling (i.e., Mux B is always a single input). This might be useful when scanning a series of dynamic inputs and a reference input such as a cold-junction compensation temperature.[11]

Three simple examples may help clarify the concept of interleaved sampling, channel scanning, and the combination of interleaved sampling and channel scanning.

Example 1: Basic Interleaved Sampling

Assume that we would like to alternately sample first channel 2 (i.e., AN2) and then channel 14 (AN14), that the channel 2 input is to be referenced to V_{REF-}, that the channel 14 input is to be referenced to AN1 (unipolar differential mode), and that we would like to have the ADC module generate an interrupt after three samples of each channel have been captured (a total of six samples per interrupt). The conversions are to be made over the full input voltage range—i.e., using AV_{SS} and AV_{DD} as the lower and upper reference voltages for the converter circuitry (as opposed to the reference voltage for the S/H circuitry). In that case,

CH0SA = 0x02	(Mux A = channel 2)
CH0NA = 0	(Mux A is referenced to V_{REF-})
CH0SB = 0x0E	(Mux B = channel 14)
CH0NB = 1	(Mux B is referenced to AN1)
ADCHS = 0x1E02	(CH0NB in Bit 12, CH0SB in <11:8>, CH0NA in Bit 4, CH0SA in <3:0>)
ALTS = 1	(Enable alternate sampling)
BUFM = 0	(Don't split ADCBUF)
SMPI = 5	(Number of samples per interrupt = 6)
CSCNA = 0	(Don't scan inputs)
VCFG = 0	($ADV_{REFH} = AV_{DD}$, $ADV_{REFL} = AV_{SS}$)
ADCON2 = 0x0019	(VCFG in <15:13>, CSCNA in Bit 13, SMPI in <5:2>, BUFM in Bit 1, ALTS in Bit 0)
ADCSSL = 0x0000	Not scanning, so this is really a "don't care" value

Example 2: Channel Scanning

Assume that we want to scan channels 1–3, 6, 8, 10, 12, and that we want to generate an interrupt after capturing two samples of each channel (i.e., after a total of 14 samples). Data will be converted over the range V_{REF-} to V_{REF+}. For this scenario, we configure the ADC module as follows:

CH0SA = 0	(Mux A = 0, although this is really a "don't care" value)
CH0NA = 0	(Mux A is referenced to V_{REF-} – "don't care" value)
CH0SB = 0	(Mux B = 0 – "don't care" value)
CH0NB = 0	(Mux B is referenced to V_{REF-} – "don't care" value)
ADCHS = 0x0000	(CH0NB in Bit 12, CH0SB in <11:8>, CH0NA in Bit 4, CH0SA in <3:0> – all are "don't care" values)
ALTS = 0	(Disable alternate sampling)

BUFM = 0   (Don't split ADCBUF)

SMPI = 13   (Number of samples per interrupt = 14)

CSCNA = 1   (Enable channel scanning)

VCFG = 3   (ADV$_{REFH}$ = V$_{REF+}$, ADV$_{REFL}$ = V$_{REF-}$)

ADCON2 = 0xC434   (VCFG in <15:13>, CSCNA in Bit 10, SMPI
                         in <5:2>, BUFM in Bit 1, ALTS in Bit 0)

ADCSSL = 0x154E   Enable <12>, <10>, <8>, <6>, <3:1> to sample
                         channels 1–3, 6, 8, 10, and 12

Example 3: Combined Interleaved Sampling and Channel Scanning

Now assume that we want to scan channels 2, 5, and 7, alternate that group with
sampling channel 4, and that we want to generate an interrupt after capturing four
samples of each channel (i.e., after a total of 16 samples). Data will be converted
over the range V$_{REF-}$ to V$_{REF+}$, and the negative input to the S/H circuit will be
set to V$_{REF-}$. For this scenario, we configure the ADC module as follows:

CH0SA = 0   (Mux A = 0, although this is really a "don't care" value)

CH0NA = 0   (Mux A is referenced to V$_{REF-}$)

CH0SB = 4   (Mux B = 0 – "don't care" value)

CH0NB = 0   (Mux B is referenced to V$_{REF-}$)

ADCHS = 0x0400   (CH0NB in Bit 12, CH0SB in <11:8>,
                         CH0NA in Bit 4, CH0SA in <3:0>)

ALTS = 1   (Enable alternate sampling)

BUFM = 0   (Don't split ADCBUF)

SMPI = 15   (Number of samples per interrupt = 16)

CSCNA = 1   Enable channel scanning

VCFG = 3   (ADV$_{REFH}$ = V$_{REF+}$, ADV$_{REFL}$ = V$_{REF-}$)

ADCON2 = 0xC43D   (VCFG in <15:13>, CSCNA in Bit 10, SMPI in
                         <5:2>, BUFM in Bit 1, ALTS in Bit 0)

ADCSSL = 0x00A4   Enable <7>, <5>, and <2> to sample channels
                         2, 5, and 7

To recap, at this point we've configured the I/O pins for our application's particular requirements, selected the voltage reference sources, and identified the ADC
input channels that are to be converted. We have two more steps left in configuring
the signal chain: specifying the ADC conversion clock and selecting the conversion
trigger.

Step 1D – Specifying the ADC Conversion Clock

Each complete digitization cycle (the acquisition and conversion of a single channel) requires one cycle of the ADC clock (whose period is designated T_{AD}) for acquisition of the analog signal level, an additional ADC clock for each bit of the conversion, and one more cycle of the ADC clock to transfer the digitized value to the ADCBUFx location and set up for the next digitization. This means that 10-bit ADCs require 13 T_{AD} and 12-bit ADCs need 15 T_{AD} to digitize a single channel. As with so much of the dsPIC DSC, the clock source and period is software configurable, but the chip's hardware requires that T_{AD} be at least 333.3 ns[12] when operating with a supply voltage V_{DD} of at least 4.5V or at least 666.7 ns when operating V_{DD} below 4.5V. Failure to meet this timing requirement will result in ADC readings that are significantly lower than they should be because the sample-and-hold amplifier will not have sufficient time to charge between channels.

T_{AD} is derived from T_{CY}, the system clock period, using the equation:

$$T_{AD} = T_{CY} * (0.5 * (ADCS<5:0> + 1)) \qquad \text{Equation 3.1}$$

where

T_{AD}	is the ADC clock period,
T_{CY}	is the system clock cycle period, and
ADCS<5:0>	is the numeric value of the 6-bit ADC Clock Select register ADCS.

From this, we can calculate that the maximum sampling rate at a V_{DD} of 4.5V or above is:

$$f_{smax} = 1 / (15 * T_{ADmin}) = 1 / (15 * 333.3 \text{ ns}) = 200,020 \text{ sps}$$

Operating at a V_{DD} of less than 4.5 V, f_{smax} is half that value or 100,010 sps.

To determine the correct value to program into the ADCS register, we simply solve for ADCS in the equation above, which yields:

$$ADCS<5:0> = 2 * (T_{AD} / T_{CY}) - 1$$

For example, if we wanted to sample all 16 channels at 10 Ksps per channel (a cumulative sampling rate of 16 × 10 Ksps or 160 Ksps), we will require a total of 160 Ksps × 15 = 2,400,000 ADC clock cycles / sec. This equates to a T_{AD} of 1 / (2,400,000 cycles/sec) = 416.7 nsec. Assuming that the dsPIC DSC is operating at its maximum 30 MIPS speed, the instruction cycle time is T_{CY} = 1 / 30 MIPS = 33.33 ns. We would then calculate the ADCS value as:

$$ADCS = 2 * (416.7 \text{ ns} / 33.3 \text{ ns}) - 1 = 24$$

It's important to note that, since the ADCS register supports only six bits, it is limited to a maximum value of 63. Thus, for a dsPIC DSC running at 30 MIPS performing 12-bit conversions, the maximum value for T_{AD} is:

$$T_{ADmax} = 33.33 \text{ ns} * 0.5 * (63 + 1) = 1.07 \text{ μs}$$

which corresponds to a minimum sampling frequency of:

$$f_{smin} = 1 / (15 * T_{ADmax}) = 1 / (15 * 1.07 \text{ μsec}) = 62,507 \text{ sps}$$

If the algorithms used by the application require a slower sampling rate, the application must either decimate the sampled values[13] or it must use a combination of polled and interrupt-driven operation to initiate a single group of digitizations for all channels that are performed individually at the higher rate but that stop after all channels have been sampled once. Some other mechanism, perhaps a software timer, is used to generate the next batch of digitizations at the desired slower rate.

This last approach is not quite as bad as it might first appear because, although it incurs additional software overhead, this method allows us to sample all of the channels relatively simultaneously, a feature that is of great benefit to systems in which signals on one channel may be related to another channel. For instance, robotic arms frequently have what's known as an *end-of-arm force/torque sensor* that measures the force and torque experienced by the object at the end of the robotic arm. This allows the robot to safely move the object without either flinging it off into space (if the robot doesn't apply enough holding force) or crushing it (if the robot applies too much force), and the force and torque are usually measured in three dimensions. A very common type of force/torque sensor is one in which the loads on three precisely positioned physical beams on the sensor are measured, so the signals from all three beams are related. If the signals are measured with too much of a time difference between the samples from each channel, the system loses information because samples from one signal apply to the state of the system at one point in time while the samples from another signal apply to its state at a slightly different time. The goal is to measure all signals as simultaneously as possible in order to get the most accurate reading. By reducing the interchannel sampling time, we can significantly improve the system's performance because the samples within a group become "more simultaneous."

Step 1E – Selecting the Sampling and Conversion Triggers

The final step in configuring the ADC Module is the selection of the sampling and conversion triggers. As we discussed previously, each digitization cycle consists of a sampling phase in which the S/H amplifier charges to the current analog signal level on its input and of a conversion phase in which the A/D converter circuitry converts the S/H amplifiers analog signal level to a numeric value. At the conclusion of the conversion phase, the digitized value is stored to the appropriate ADCBUF entry. The dsPIC DSC allows the user to determine how each of the two phases will start (known as the *sample trigger* and the *conversion trigger*), giving the designer a great deal of flexibility.

There are two options for the sample trigger: manual triggering, in which the application sets the sampling flag every time it wants to convert an input signal, and automatic trigger, in which sampling of the next channel is started as soon as the conversion for the current channel completes. Manual triggering is appropriate when the signal being monitored can be checked on a nonperiodic basis (for instance, background monitoring of a fluid level to make sure it doesn't overflow) or when samples need to be taken at a slower rate than can be supported in auto conversion mode. An example of the latter case would be sampling the temperature of a device with a large thermal mass;[14] because the temperature can change only relatively slowly, the application may sample it infrequently to reduce the associated software overhead. Although the sampling rate is low, it still needs to be periodic in order to correctly implement DSP algorithms. To sample periodically but slowly, the application may perform the digitization within the interrupt service routine for a timer with the desired period.

The application selects the desired sampling trigger through the A/D Sample Auto-Start (ASAM) bit in ADCON1 (ADCON1<2>). Clearing ASAM configures the ADC module to use a manual sampling trigger, in which the application must explicitly set the A/D Sample Enable (SAMP) bit in ADCON1 (ADCON1<1>) to a value of "1" every time it wants to sample the input signal. Conversely, setting ASAM configures the module for automatic sampling in which sampling of the current signal level is automatically begun as soon as the previous conversion completes, with the SAMP bit being set by the module's hardware without application intervention. In either manual or automatic sampling, the S/H amplifier will continue to sample the input signal until the conversion phase begins, based on the occurrence of the conversion trigger.

The application identifies the method by which the conversion phase will start (and hence the sampling phase will end) through the Conversion Trigger Source Select (SSRC<2:0>) bits in the ADCON1 register (ADCON1<7:5>). Although there are three SSRC bits and therefore eight possible conversion trigger sources, three of those combinations are reserved:

SSRC<2:0>	Conversion Trigger Source
000	Clearing SAMP bit ends sampling and starts conversion
001	Active transition on INT0 pin ends sampling and starts conversion
010	Timer 3 compare event ends sampling and starts conversion
011	Expiration of the motor control PWM interval ends sampling and starts conversion
100	Reserved
101	Reserved

110	Reserved
111	Internal counter ends sampling and starts conversion (auto convert)

Table 3.1. Conversion Trigger Source Bit Mapping for ADCON1 SFR

Note that once the SSRC bits have been set, they remain in effect until the chip either powers down or they are overwritten by a subsequent write to ADCON1.

When using manual conversion, the application is responsible for terminating the sampling phase and starting the conversion phase by simply first setting SSRC to 000b and then clearing the SAMP bit. The application waits for the conversion to complete by either delaying for an appropriate length of time or by monitoring the A/D Conversion Status flag (DONE) in the ADCON1 register (ADCON1<0>), which will be set once the conversion has finished. Subsequent samples are taken by repeating the SAMP strobe/DONE monitor sequence.

The designer can also start the conversion phase through the occurrence of one (and only one) of three events: an active transition on the INT0 signal (an external source), the expiration of the dsPIC DSC's Timer 3 (an internal source), or the conclusion of the chip's motor control PWM interval (also an internal source). In order to use this mode, the application must set the SSRC field to the desired trigger source (either 001b, 010b, or 011b) and configure the corresponding resource (external interrupt, timer, or PWM) appropriately. Once the selected event occurs, the module stops the current signal sampling and begins the conversion phase.

Finally, the application can use the *auto convert mode*, in which the dsPIC DSC automatically starts the conversion phase a specified number of A/D clock cycles (between 1 and 32) after the start of the sampling phase. In this mode, the module uses an internal counter to monitor the length of the sampling phase and then starts the conversion phase once the specified length of time (expressed in T_{AD}) has elapsed. By combining automatic sampling with automatic conversion, the application can run the system in a fully automatic mode that requires no user intervention once it starts.

Checklist for Using the ADC Module

To perform a conversion using the ADC module, the application must perform the following seven basic operations:

1. Configure the ADC module for the application's specific signal path requirements, including:
   - configuring the associated I/O pins as analog inputs, selecting the appropriate voltage reference signals, and configuring any of the other I/O port signals as digital inputs or outputs;

- selecting the ADC input channels that are to be converted;

- selecting the desired ADC conversion clock and trigger; and

- enabling the ADC.

2. Configure the ADC interrupt, if the application is performing interrupt-driven data acquisition, by
   - clearing the A/D Interrupt Flag (ADIF) and

   - setting the ADC interrupt priority.

3. Begin sampling the analog data.
4. Wait for the necessary data acquisition time.
5. Trigger the end of the data acquisition and start the data conversion.
6. Wait for the data conversion to finish by either:
   - waiting for the ADC interrupt, or

   - monitoring the ADC Done flag to see when it is set.

7. Read the converted value(s) from the ADC Result Buffer (ADCBUF0: ADCBUF15) and clear the ADIF flag if performing interrupt-driven acquisition.

Timer/Counter Module

The timer/counter module allows the dsPIC DSC to generate accurate internal clocks that can be used to maintain a real-time clock (RTC), to track elapsed time for application-specific events, to schedule time-critical events such as ADC sampling triggers, or to count transitions on certain external hardware signals. Depending on the model of dsPIC DSC used, the module supports either three or five individual timer/counters that can operate independently or in specific predefined combinations to offer a tremendous amount of flexible timing and event counting power.

There are three basic timer/counter configurations that may be present in a particular dsPIC timer module, denoted in the documentation as Type A, Type B, and Type C timer/counters. All of these three types share certain common characteristics, but they each also have unique features that allow them to perform certain specific tasks very well. In our examination of the module, we'll first look at the aspects common to all dsPIC timer/counters and then explore the distinctive features of each type to understand how and why a designer would want to incorporate a particular configuration into an application. Note that in the following discussion we'll frequently refer to register or pin names such as PRx, TMRx, TxCON, etc.

In these cases, simply substitute the desired timer index (1–5) for the "x" to get the actual name of the register or pin. For example, the registers for Timer 1 are PR1, TMR1, T1CON, while those for Timer 2 are PR2, TMR2, and T2CON.

Common Timer/Counter Features

Every timer/counter has the following minimum set of features:

1. a 16-bit Period register (PRx),
2. a 16-bit counter (TMRx),
3. a 16-bit comparator that compares the value in the counter to the value in the Period register on every clock cycle,
4. a clock prescaler to control the frequency at which the timer/counter is updated,
5. two or more timer/counter clocking options, one of which is always the instruction cycle clock (T_{CY}),
6. a Gated Timer Accumulate option that allows the application to measure the length of time that the associated Timer Clock input pin was pulsed high or low,
7. the ability to set an interrupt flag and to generate an interrupt when the comparator detects that the counter value matches the value stored in the Period register, and
8. a Timer Configuration (TxCON) register that allows the application to configure the operation of the specific timer.

The basic operation of a timer is really very simple: the application clears the timer's counter register (TMRx), loads a target value in the timer's Period register (PRx), and then enables the timer by setting the Timer On (TON) flag in the timer's control register (TxCON). Once enabled, the timer increments its Counter register on each rising edge of the selected clock source, and when the Counter register value matches the Period register value, the timer sets its interrupt flag and generates an interrupt (assuming the interrupt is enabled). Setting the interrupt flag also resets the Counter register to 0 without software intervention, and the process repeats until the user disables the timer by clearing its TON flag.

Although correct insofar as it goes, the preceding summary of the basic timer operation glosses over several important concepts that one needs to understand in order to configure the timer properly. The first of these two concepts is the selection of an appropriate clock source for the timer/counter, the second is the operation of the clock prescaler, and the third is the calculation of the desired value to load into the Period register, which depends upon both the selected clock source and the clock prescaler. We'll turn our attention to these three related subjects now.

The application can configure each timer to use either an internally generated clock based on the instruction cycle clock (with a period of T_{CY}) or an external clock signal (with a period of T_{XCK}) that is applied to the TxCK pin. In either case, the selected clock signal passes through a prescaler that divides it by a factor of either 1, 8, 64, or 256 (i.e., the clock period is multiplied by a factor of 1, 8, 64, or 256, respectively) prior to its use by the associated counter register. This gives the designer the ability to track longer time periods at the expense of a lower time resolution, as we can readily see from a simple example. If we choose to use the internal instruction cycle clock of a dsPIC DSC running at 30 MIPS ($T_{CY} = 33.33$ ns) as our clock source and a prescale factor (PSF) of 1, we can achieve a timer resolution of:

$$T_{RES} = T_{CY} * PSF = 33.33 \text{ ns} * 1 = 33.33 \text{ ns,}$$

and our 16-bit Counter register will support any period up to:

$$T_{MAX} = 65,536 * 33.33 \text{ ns} \approx 2.18 \text{ ms.}$$

If, on the other hand, we use the same clock source but employ a prescale factor of 256, our timer resolution is reduced to:

$$T_{RES} = T_{CY} * PSF = 33.33 \text{ ns} * 256 = 8.5325 \text{ μs,}$$

but the 16-bit Counter register can now support an period up to:

$$T_{MAX} = 65,536 * 8.5325 \text{ μs} \approx 559.18 \text{ ms.}$$

Once the clock source and prescaler values have been chosen, the value to load into the period register is easily computed as:

$$PR = T_{PERIOD} / T_{RES}$$

where

PR	is the 16-bit value to load into the Period register (PRx)
T_{PERIOD}	is the desired period in seconds
T_{RES}	is the timer resolution in seconds

Of course, T_{PERIOD} and T_{RES} must be such that the Period register value fits into a 16-bit register—i.e., PR must be less than or equal to 65,535. In many applications, it's the need to meet this register-length requirement that drives the choice of the clock source and the prescaler value.

Now that we've looked at the traits that all three timer/counter types share, let's examine the ways in which they differ and why we would want to use one type of timer/counter over another in particular situations.

Type A Timer/Counters

Type A timer/counters can be used either as a 16-bit timer (as we've already discussed), as a 16-bit synchronous counter, or as a 16-bit asynchronous counter. When operating as a 16-bit synchronous counter, the timer/counter behaves in much the same way as when it serves as a 16-bit timer; the only real difference is that, whereas in timer mode one expects the clock to be stable and periodic, in counter mode the input signal to be counted (which comes in on the associated external clock pin) may occur on an aperiodic basis. The timer's hardware synchronizes the count signal to the internal phase clocks before feeding the synchronized clock to the counter, which is incremented on each rising edge of the external count signal. As in timer mode, when the TMR1 register value equals the value of the Period register PR1, the TMR1 register is reset to 0 and the module generates an interrupt.

The asynchronous counter mode is unique to Type A timer/counters, as is the ability to use a 32-kHz low-power oscillator for its external clock signal, and these two features make it ideal for use as a real-time clock (RTC) for dsPIC applications because they permit the timer to operate even when the system clock is inactive. This means that timer will still work when the chip is in Sleep mode, a critical requirement for a true RTC. In any dsPIC DSC, there is only one Type A timer, Timer 1. Figure 3.12 shows the block diagram for the Type A timer/counter.

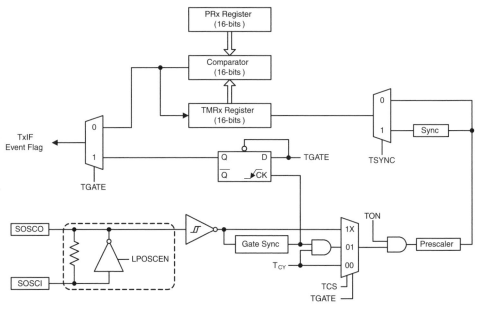

Figure 3.12. Type A Timer/Counter Block Diagram

There are several important points to note in the block diagram. First, because Timer 1 has special circuitry across its Timer 1 Clock Input (T1CKI) and its Secondary Oscillator Input (SOSCI) signals, an application can use an external low-power 32-kHz crystal to generate the clock for the timer. Second, whatever clock source is used, the clock signal passes through the prescaler before being optionally synchronized to the internal instruction cycle clock, so that input clocks with frequencies that are at least 0.5 * (F_{CY} * *PSF*) can be used (i.e., input clocks can be faster than the instruction cycle clock by a factor of ½ * *PSF*). However, because of the timing constraints of the combinatorial logic used to implement the asynchronous external clock input, external clock signals to a Type A timer/counter always must be less than 25 MHz for proper operation, even if this limits the frequency range that might be expected by applying the prescaler factor. Of course, as with all types of dsPIC timer/counters, the external clock must meet the required rise and fall times, which are a maximum of 10 ns.

Now that we know what a Type A timer/counter can do, what are the specific steps that we need to take to configure it for operation? First, we have to determine the mode in which we wish to operate the timer/counter: timer, synchronous counter, or asynchronous counter (the mode used to implement a real-time clock). Having established the operational mode, we then need to determine when (or even if) we want the timer/counter to generate an interrupt (i.e., on which counter value should the timer/counter interrupt the chip?).

Timer Mode Initialization

To initialize Timer 1 for operation in Timer mode,

1. Stop the timer during initialization by clearing the Timer On (TON) control bit in T1CON (T1CON<15>). Also, clear the Timer 1 Interrupt Flag (T1IF) in IFS0 and the Timer 1 Interrupt Enable (T1IE) flag in IEC0 to clear out any pending Timer 1 interrupt conditions. This ensures that the initialization sequence can complete without interruption.

2. Clear the Timer Clock Select (TCS) flag in T1CON (T1CON<1>) to select the internal instruction cycle clock for the input to the timer's prescaler.

3. Set the clock prescaler to the desired value by configuring the Timer Clock Prescale (TCKPS) bits in T1CON (T1CON<5:4>):

    TCKPS = 00b   use 1:1 clock prescaler (effectively no prescaling)

    TCKPS = 01b   use 1:8 clock prescaler

    TCKPS = 10b   use 1:64 clock prescaler

    TCKPS = 11b   use 1:256 clock prescaler

4. Clear the Timer External Clock Input Synchronization Select (TSYNC) flag in T1CON (T1CON<2>), since we are using the internal clock and thus don't need to synchronize the external clock signal.

5. Set the Timer Stop in Idle (TSIDL) flag in T1CON (T1CON<13>) based on whether the application wants the timer to stop in Idle mode or to run in Idle mode:

TSIDL = 0   allow the timer to run in Idle mode

TSIDL = 1   stop the timer when the chip enters Idle mode

6. Set the Timer Gated Time Accumulation Enable (TGATE) flag value in T1CON (T1CON<6>) based on whether the application is using the gated time accumulation mode:

TGATE = 0   disable gated time accumulation mode

TGATE = 1   enable gated time accumulation mode (must also use the internal clock source in this mode, so TCS must be cleared if TGATE = 1)

7. Set the Period register (PR1) to the value that corresponds to the desired elapsed time between interrupts, as previously calculated in the *Common Timer/Counter Features* section.

8. If the application needs to be interrupted upon expiration of the timer (i.e., when the TMR1 value equals the value of PR1 set in step 7 or on the falling edge of the gating signal if gated time accumulation mode is enabled), then set the Timer 1 Interrupt Enable flag (T1IE) in IEC0 to enable the interrupts. Also enable global interrupts, if that has not already been done.

9. Start the timer by setting TON in T1CON.

Note that in the initialization, steps 2 through 6 are usually combined into a single write to the T1CON register; they are broken out in the initialization sequence only for clarity.

Synchronous Counter Mode Initialization

The synchronous counter mode is similar to the timer mode except that the signal to be counted is applied on the external clock input and is synchronized to the internal clock. In this mode, the gated time accumulation concept does not apply. The initialization sequence, therefore, is very similar to that for timer mode operation.

1. Same as step 1 in the *Timer Mode Initialization* section.

2. Set the Timer Clock Select flag in T1CON to a value of 1 to select the external clock input.

3. Same as step 3 in the *Timer Mode Initialization* section.

4. Set the Timer External Clock Input Synchronization Select (TSYNC) flag in T1CON to 1 to synchronize the external clock input to the instruction cycle clock.

5. Same as step 5 in the *Timer Mode Initialization* section.

6. Clear the Timer Gated Time Accumulation Enable (TGATE) flag value in T1CON since we are not using the gated time accumulation mode.

7. Same as step 7 in the *Timer Mode Initialization* section.

8. Same as step 8 in the *Timer Mode Initialization* section.

9. Same as step 9 in the *Timer Mode Initialization* section.

As with the timer mode initialization, steps 2 through 6 are usually combined into a single write operation to the T1CON register.

Asynchronous Counter Mode

Asynchronous counter mode is identical to synchronous counter mode except that the counter will continue to operate in Sleep mode and the input can be a low-power 32-kHz crystal oscillator. If a crystal oscillator is used as the input, one side of the oscillator is connected to the external clock input signal and the other side is connected to the Secondary Oscillator Input pin (SOSCI). In addition, the Low-Power Oscillator Enable (LPOSCEN) flag in the Oscillator Control (OSCCON) register has to be set to a "1" to enable the crystal oscillator support circuitry required to drive the crystal.

As would be expected, the initialization sequence is very similar to that for the synchronous counter.

1–3. Same as steps 1 through 3 in the Synchronous Counter Initialization section.

4. Clear the Timer External Clock Input Synchronization Select (TSYNC) flag in T1CON (T1CON<2>) to operate the external clock signal asynchronously.

5–9. Same as steps 5 through 9 in the Synchronous Counter Initialization section.

As with the other initializations, steps 2 through 6 are usually combined into a single write operation to the T1CON register.

Type B Timer/Counters

Like their Type A counterparts, Type B timer/counters may serve as 16-bit timers or 16-bit synchronous counters; however, they also have the ability to be combined with a Type C timer/counter to form a 32-bit timer or a 32-bit synchronous counter. When combined in this manner, the Type B timer/counter forms the lower 16 bits of the 32-bit value, and the Type C timer/counter forms the upper 16 bits. This added flexibility comes at a small price, though, because Type B timer/counters cannot function asynchronously in any capacity, nor can they accept a crystal oscillator input. In most applications, these are not serious constraints.

On devices that support Type B timer/counters, Timers 2 and 4 (if present) are always of this type. In addition, when pairing Type B and Type C timer/counters, only certain combinations are permitted. Timer 2 is always combined with Timer 3, and Timer 4 is always combined with Timer 5. When combined as a 32-bit unit, the Timer Control register (TxCON) for the Type B timer/counter serves as the control register for the entire 32-bit timer/counter. Since the dsPIC DSC cannot read/write a single 32-bit value directly from/to the combined TMR3/TMR2 or TMR5/TMR4 registers, the chip uses a holding register (TMRxHLD) for values in the upper word of the timer/counter register. To write a 32-bit value to the combined timer/counter register, the application first writes the upper 16 bits to the TMRxHLD register and then writes the lower 16 bits to TMRx for the Type B timer/counter. Upon receiving the write to TMRx, the dsPIC DSC simultaneously loads the value of TMRxHLD into the TMRx register for the corresponding Type C timer/counter. Reading the 32-bit value from the combined counter register is reversed; when the application reads the lower word from the Type B counter register, the dsPIC DSC automatically loads the upper word into the TMRxHLD register. The application can then read the TMRxHLD register to get the upper word value.

Figure 3.13 shows the block diagram for a Type B timer/counter. The only significant differences between this block diagram and that for the Type A timer/counter is the absence in the Type B block diagram of the signal-conditioning circuitry that supports crystal oscillator input and the lack of an asynchronous clock path from the output of the prescaler into the counter circuitry. As with the Type A timer/counter, the prescaler for the Type B timer/counter precedes the synchronization circuitry, so Type B timer/counters can accept external clock signals up to $0.5 * F_{CY} * PSF$. Because there is no asynchronous combinatorial logic on the external clock signal input, Type B timer/counter clock signals are not limited to the 25-MHz maximum frequency required by Type A circuits.

Type B timer/counters have one additional flag defined in their TxCON register that doesn't appear in the corresponding register for either Type A or Type C timer/counters. The application should set the 32-bit Timer Mode Select (T32) bit to combine a Type B timer/counter with the corresponding Type C timer/counter or clear the flag to operate the two timer/counters as independent 16-bit units. Because Type B timer/counters always operate synchronously, there is no need for the TSYNC bit in their TxCON registers, so that flag is not defined.

16-bit Timer and 16-bit Synchronous Counter Initialization
Initialization of a Type B timer/counter for operation either as a 16-bit timer or as a 16-bit synchronous counter is identical to the corresponding initialization sequence for a Type A timer/counter except that T32 must be cleared in TxCON and TSYNC is not initialized (since it doesn't exist for Type B units).

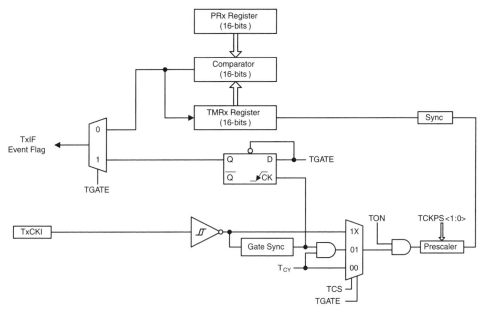

Figure 3.13. Block Diagram of a Type B Timer/Counter

32-bit Timer or Synchronous Counter Initialization

Initialization of a Type B timer/counter for 32-bit operation is identical to that for operation as a 16-bit timer, except that T32 must be set in TxCON, and the interrupt enable flag that must be used is that for the Type B timer/counter. Note that the PR register for both the Type B timer/counter and its paired Type C timer/counter must be written.

Type C Timer/Counters

Type C timer/counters are similar to Type B timer/counters and have the same basic operational modes, but as the block diagram in Figure 3.14 shows, the external clock synchronization circuitry is placed before the prescaler rather than after, so the maximum external clock frequency for a Type C timer/counter is $0.5 * F_{CY}$ rather than the potentially higher $0.5 * F_{CY} * PFS$ of Type B timers. Like Type B timer/counters, the lack of asynchronous combinatorial logic on the external clock input eliminates the 25-MHz clock frequency limitation of Type A units.

Initialization of the Type C timer/counters is nearly identical to that for the Type B timer/counters in each of the four different operating modes, with one important difference. When operating a Type C timer/counter in 16-bit mode, the T32 flag in its paired Type B timer/counter control register must be cleared to place both timer/counters in 16-bit mode.

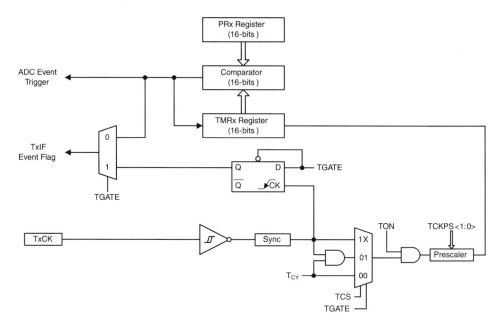

Figure 3.14. Block Diagram of a Type C Timer/Counter.

3.4   Summary

This chapter has presented quite a bit of detailed information about the dsPIC DSC's architecture, although it has certainly not explored all there is to understand about the chip. Additional detailed information can be found in three other Microchip documents: the *dsPIC30F Family Reference Manual,* the *dsPIC30F Programmer's Reference Manual,* and the data sheet for the application's specific dsPIC device. All of these can be found on the official Microchip website (*www.microchip.com*), as can numerous application notes, design tips, and web seminars that describe and analyze particular system components and their application to real-world problems.

In the next chapter, we'll continue our exploration of the peripheral modules available on the dsPIC DSC by delving into the chip's communication resources.

Endnotes

1. The requirement for *deterministic* system behavior simply means that the design must guarantee that the system will respond to an event within a specified time period, with the further implied constraint that the response time is suitable for the tasks that we're trying to accomplish. It does no good to ensure that the system responds to an event within three seconds if new events occur every three milliseconds.

2. *Errata* is the plural form of the Latin word erratum, which is the term used in the Latin formula for the assignment of mistakes in a court case. In our case (no pun intended), one gathers a bit more insight from the verb errare (to stray) from which "erratum" is derived. The errata are a list of the conditions under which the chip's implementation strays from its intended design. Definitions come from *The American Heritage Dictionary* entry and the *Thompson-Gale Legal Encyclopedia* entry found on the URL *http://www.answers.com/topic/erratum*.

3. When we *sign-extend* a two's complement value, we replicate the sign bit as necessary to reach the desired representation length. For example, if we wanted to sign-extend an 8-bit value to 16 bits, we would replicate the sign bit (the MSB) 8 times to the left of the number. Sign-extending the 8-bit value 0xFC to 16 bits would result in the value 0xFFFC; sign-extending the 8-bit value 0x7F would result in the value 0x007F.

4. The *signal chain* refers to the processing stages that a signal goes through from input to output. This chain usually consists of both analog and digital components, at least in systems that employ digital signal processing techniques.

5. psi = pounds per square inch.

6. *dsPIC30F 12-bit ADC Module, Part 1 of 2*, page 6. The webinar is available from the Microchip website.

7. Actually, the term *conversion* is one that is used for not only the act of converting the analog signal level of the sample-and-hold amplifier (the manner in which we use it in this section) but also for the overall acquisition/conversion cycle. In most cases, the intended meaning is fairly obvious from the context, but it's an example of how sloppy terminology can work its way into the engineering lexicon.

8. *Jitter* refers to the deviation about a nominal value of the sampling period. For instance, if the nominal sampling period for the system is 2 ms and the actual sampling period varies between 1.8 ms and 2.1 ms, the jitter for the system would be 0.3 ms or 15%, which is not too good.

9. Yes, those really are numeric digits at the front of the two "words," if one can call an amalgamation of digits and letters a word. The key is to remember the convention, not to get hung up on the rules of English. Although this semi-mnemonic is not original with the author, it has proven useful over the years.

10. *Common-mode noise* is electrical noise that appears on both inputs of a differential signal. This commonly appears on twisted-pair cables in which both conductors are intertwined and thus any radiated noise that is coupled into one wire will be coupled into the other as well. Theoretically, the differential signal is immune to this type of noise because the receiver looks at just the difference between the voltages on the two conductors, so any voltage (including noise voltage) that appears on both conductors is subtracted out. In practice, this is not always the case, but it is still a very useful technique for significantly reducing the effects of unwanted noise.

11. *Cold-junction compensation* is a technique used to account for the temperature-dependent voltage differential that develops at the junction of any two dissimilar metals. We'll examine the topic in detail in the multichannel temperature-monitoring application developed later in the book.

12. 1 ns = 1 nanosecond = 10^{-9} seconds, which the reader already knew. Having just ranted about sloppy terminology, however, it seemed wise to leave nothing to chance.

13. *Decimation* is the process by which data sampled at one frequency is converted to a sampling rate at a lower frequency. When the higher sampling rate is an integer multiple of the lower frequency, this can be easily accomplished by simply ignoring samples in the higher-rate data stream whose sample index is not an integer multiple of the lower-rate stream. If the higher sampling rate is not an integer multiple of the lower rate, however, we must first *interpolate* the higher frequency data to generate a third data stream whose sample rate is the least common multiple of the original high sample frequency and the desired lower sample rate. We can then safely decimate this third data stream to extract the desired lower-rate data.

14. *Thermal mass* is a concept in thermal management that is similar to physical mass in mechanics. The greater a body's thermal mass, the more difficult it is to change its temperature. An example of an object with a high thermal mass would be a large plate of steel, which requires a great deal of energy to heat but that stays warm for a long time once it is hot. A thin needle is an object with a relatively low thermal mass; it can be heated and cooled quickly.

Learning to be a Good Communicator

*The higher you go, the wider spreads the network of communication
that will make or break you.*

—Donald Walton

One aspect that permeates everything we do with intelligent sensors, indeed one of the defining qualities of an intelligent sensor, is its ability to communicate with other components both within its local environment and remotely. As we'll see, the dsPIC DSC offers an extensive array of communication options, with the appropriate method dependent upon the nature of the data to be communicated, the speed at which the data must be transferred, and the physical medium over which the communications must take place. The chapter is divided into two main sections, the first of which provides background on the types of communication requirements that frequently arise in intelligent-sensing applications. The remainder of the chapter explores the communication options available for the dsPIC family of devices, identifies the situations for which each option is most appropriate, and explains specifically how to use the associated hardware modules.

4.1   Types of Communications

As we examine the issue of communication, it's important to understand that there is no one "right" form of communication for intelligent sensors any more than there is a single best form of communication for people. When we want to get a point across to someone else, we instinctively adjust the volume of our voice, the speed of our speech, and the amount of information we're trying to convey based on the environment, the importance of what we have to say, and the person to whom we're speaking. If one were wandering the halls of the Louvre with a four-year-old niece, the phrase "pretty picture" might be all the commentary required; the same excursion with a fellow painter might elicit exhaustive debate on brush stroke, color, and style. In both cases, the conversation is shaped by the capabilities of the speaker and the audience, both in terms of the amount of information that is communicated and its level of detail.

The situation is no different with intelligent sensors; the methods of communication we use depend upon the type of data that we want to communicate, its urgency, and the physical medium it must traverse between the source and destination. Configuration commands, for instance, usually are sent from the host system to the sensor system before the sensor starts reporting parameter measurements. Unless the quantity of command data is extremely large, the speed at which that data is transferred is not particularly important, and other than the requirement that commands arrive at the sensor in the order in which they were transmitted, the communications are not particularly time-critical. At the opposite end of the spectrum would be alarm conditions detected by the sensor; these must reach their destination reliably and quickly enough that the host system can act upon the received information to avert undesirable, possibly dangerous, operational conditions. We would expect the latter situation to demand a communication technique that is more robust and timely than the former. Fortunately, the dsPIC DSC offers a number of options that meet the needs of a wide variety of communication situations.

Defining Characteristics of a Communication Channel

While we've touched upon some of the important characteristics of a communication channel, it's time that we become more explicit. Doing so allows us to better categorize different approaches and gives us some benchmarks with which to assess alternative communication options. In our discussion, we're making the assumption that we'll be using wired electronic communication channels (as opposed to optical, wireless, or some other media). Before you dismiss this as an "of course" kind of assumption, be aware that there are plenty of systems that use either optical or wireless communications; we make the assumption here simply because we're exploring the options available on the dsPIC DSC.

The key criteria that we'll use when making our communication decisions are:

1. the required sustained (as opposed to burst) channel throughput, expressed in bits per second;

2. whether the link is point-to-point (i.e., between only two devices) or whether it is a multidrop link (i.e., more than two devices on the same link);

3. the physical length and the electrical noise level of the communication link;

4. whether the communication link needs to be synchronous, asynchronous, or can be either;

5. the need for hardware-based error detection.

Let's look at each of these criteria in detail before exploring the communication options available to us on the dsPIC DSC.

Channel Data Throughput

One of the most important communication-channel characteristics is the channel's *data throughput*—i.e., the amount of data that it can transfer in a given period of time. There are two principal types of data throughput, *burst throughput* and *sustained throughput*, and it's important both to understand the difference between the two terms and to know which one is the constraining factor for a particular application. Burst throughput is the fastest data rate that the channel can support for small quantities (bursts) of data. Knowing the burst throughput rate by itself is insufficient; the designer must also know the maximum amount of data that can be transmitted at that rate. While a channel's burst throughput may be exceptional, that mode of operation may not be an option if it can't transfer large enough packets of data to meet the needs of the specific situation. In contrast, the *sustained throughput* is the maximum data rate at which the channel can communicate data continuously. This is generally a better measure of the channel's data-carrying capacity, although it may be significantly lower than the burst throughput rate because the sustained rate reflects a guaranteed transfer speed that is independent of the amount of data to be sent.

In our sensor designs, we will focus on a channel's sustained data throughput, although we will note cases in which burst-mode operation may be appropriate.

Point-to-Point vs. Multinode Networks

Communication systems fall into one of two categories. Systems that employ a point-to-point topology[1] have essentially two nodes that transfer data between each other, as shown in Figure 4.1. Although it is possible to have in-line repeaters that receive from one node and transmit to another, the key is that for each link there is only one transmitter/receiver pair for each direction of transmission. This makes life relatively easy for the designer because she does not have to be concerned with scheduling data transmission access to the link since there is only one transmitter. In addition, using a point-to-point topology greatly simplifies the termination of data cables, a requirement for long links.

The primary disadvantage of point-to-point systems is their limited ability to share data with other systems. It's hard to get the word out when you're talking to just one other person, which is one reason that many systems use a *multidrop* topology in which more than two nodes are connected together via a single communication channel, as shown in Figure 4.2. With this approach, it's possible to talk to all of the nodes at once from a single node, or to establish communications between two or more nodes. Note, however, that although we can have many nodes *listening* at one time, only a single node can actually *transmit* data, which raises one of the topology's biggest challenges: how to schedule transmissions from individual nodes so that two or more nodes don't try to send data at the same time.

Point-to-Point Network

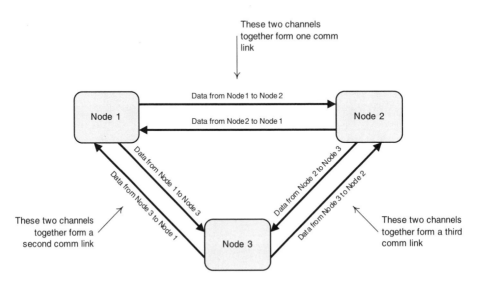

Figure 4.1. Point-to-Point Communication Network

Multi-drop Network

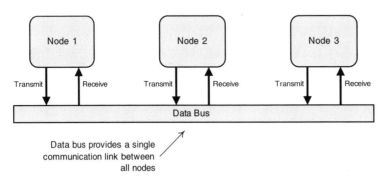

Figure 4.2. Example Multidrop Network

One of two approaches is commonly taken to overcome the difficulty of transmission *collisions*—i.e., those occasions when more than one transmitter is trying to send data at the same time. The first approach tries to avoid collisions altogether by passing around a software or a hardware *token* that tells an individual transmitter when it is safe to send data. When a transmitter has the token, it may transmit any data it has and then passes the token to the next node in the network. Should a node receive the token but have no data to transmit, it simply passes the token to the next node without taking any other action. Although this method avoids collisions, it does so at the expense of the additional software and/or hardware required to sequence the token through the network, and there may be (relatively) long periods of time in which the token is being passed among nodes with no data to send while a node that does have data to transmit waits to receive the token.

The second approach uses a somewhat different technique. Rather than trying to avoid collisions, it allows collisions to occur but detects them quickly and reliably and then tries to reschedule at least one of the offending transmitters to send its data later, presumably after the dominant transmitter (i.e., the one that gets to transmit) has finished sending its data. This can substantially improve throughput in networks in which individual nodes need to transmit data on a nonperiodic basis since it allows the nodes to attempt to transmit immediately rather than having to wait their turn. Of course, the rescheduling algorithm in the second approach needs to ensure that all nodes eventually get transmission access to avoid the possibility that a node is perpetually held off from sending its data.

Physical Properties of the Data Link

Physical properties of the data link, such as its length and the electrical noise level of the environment in which it operates, limit the maximum transfer speed and may mandate the use of special circuitry to adequately drive the electronic signals, to detect error conditions, or both. We can overcome these problems to a certain extent by using high-drive capability differential signals, but the additional power requirements and financial costs may be more than a particular application can support. As one might expect, the longer the link or the noisier the environment, the more slowly the data must be transmitted in order to be received reliably.

Asynchronous vs. Synchronous Data Transfer

In general, data can be transmitted either *synchronously* with an accompanying clock signal or *asynchronously*, without an accompanying clock signal. As always, there are trade-offs to either technique; the choice of the appropriate method is situation-dependent.

The advantage of synchronous data transfer is that the transmitter can generate a clock signal that tells the receiver precisely when to sample the incoming data in order to determine whether that data is a logical "0" or a logical "1" (assuming, of course, that we're dealing with binary data). This greatly simplifies accurate data detection at the receiving end and ensures that communications are always sequenced properly, but it comes at a heavy cost. Not only must the data link support an additional clock signal (or two, if the clock is sent differentially), but it becomes critical that the clock signal be kept very clean since any significant "glitch" on it may be interpreted erroneously as a clock transition. Even worse, the clock signal usually must have sharp edges in order to clearly identify the associated data-bit value, which generates high-frequency noise that may exceed legal limits. An example of a simple synchronous signaling scheme is shown in Figure 4.3.

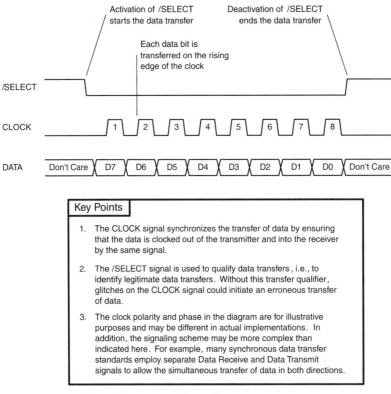

Figure 4.3. Sample Synchronous Signaling Scheme

Given all of these problems, one might be tempted to assume that asynchronous transfer is a superior approach, and often that is true. The great advantage of an asynchronous link is that it requires no external clock signal; the clock required to decode the incoming bits is contained in the data itself, neatly eliminating both the troublesome problems associated with an external clock and the cabling required for that signal. By now the reader is fully aware that such great good fortune usually comes at a price, and this is no exception. Accurate recovery of the clock embedded in the data stream requires the synchronization of a high-speed clock at the receiver (on the order of four to sixteen times that used to generate the data) with the incoming data stream, mandating special hardware and effectively restricting the upper limit of the transfer speed to a value less (often substantially less) than that achievable in a synchronous system. Figure 4.4 illustrates what an asynchronous signaling scheme might look like.

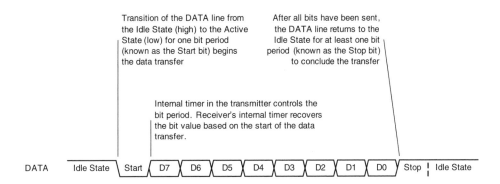

Figure 4.4. Sample Asynchronous Signaling Scheme

Hardware Error Detection

In any communication system, it is a given that errors will occur no matter what precautions are taken to avoid them. That being the case, it is incumbent upon the designer to build into the system ways in which such errors at least can be detected, if not corrected, so that erroneous data doesn't cause undesirable system operation. Depending upon the speed of the data link, it may be possible to perform this error detection in software, albeit at the expense of added processing overhead.

In some cases, however, the link speed or reliability requirements may require the use of hardware-based error detection to ensure proper communications between transmitter and receiver. Unfortunately, such hardware-based solutions tend to be both complex and expensive in terms of board space and cost unless they are incorporated into standard communication devices that can be produced in quantities high enough to amortize the development and production costs associated with the specialized circuitry. As we'll soon see, the dsPIC DSC includes at least one communication interface, the Control Area Network, that meets these criteria and offers the high degree of communication link reliability that hardware error detection can provide.

4.2 Communication Options Available on the dsPIC30F

The dsPIC DSC offers a broad variety of interfaces that allow it to communicate with other devices on the same printed circuit board or with remote systems. These interfaces include:

1. the Synchronous Peripheral Interface (SPI),
2. the Inter-Integrated Circuit (I^2C) interface,
3. the Universal Asynchronous Receiver/Transmitter (UART),
4. the Controller Area Network (CAN) interface.

We will look at the SPI, the UART, and the CAN interfaces in detail. The I^2C interface uses many of the same principles as the SPI, so it is not addressed here.

The Serial Peripheral Interface (SPI) Port

The Serial Peripheral Interface, or SPI as it is more commonly called, is a synchronous serial interface that is designed primarily to transfer data between devices that are all located on a single printed circuit board (PCB), although it can be used to communicate between PCBs as well. The interface is fairly simple, consisting of a Serial Data Out (SDO) signal, a Serial Data In (SDI) signal, a Serial Clock signal (SCK), a Chip Select (CS) signal, and a Slave Select (SS) signal. All of these signals are single-ended digital signals, one of the reasons that the SPI is not well-suited to long data links or noisy environments. Because it is so easy to implement and troubleshoot, many devices, both microprocessors and peripheral chips, employ the SPI. For example, the dsPICDEM board uses one of its two SPI ports to communicate with the on-board temperature sensor, sending configuration data to the sensor and reading temperature and status values from it.

Over the years, the SPI has evolved to support four basic modes of operation (imaginatively denoted Mode 1, Mode 2, Mode 3, and Mode 4), that operate in basically the same manner but which employ different timing relationships between the SCK clock edge and the SDO and SDI data signals to determine when to transmit data and when data is valid at the receiver. Most devices support only a subset of these modes, so it's important to make sure that both the transmitting and receiving device are able to support at least one common operating mode. Figure 4.5 shows the four possible SPI operating mode combinations. In our examples, we will use Mode 0 since it is one of the more common configurations.

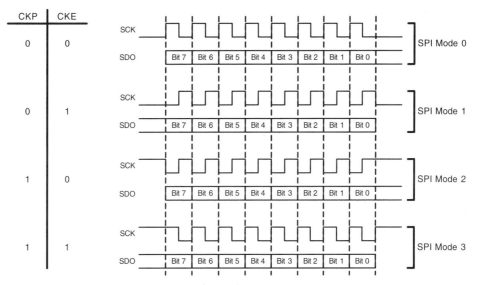

Figure 4.5. The Four SPI Operating Mode Combinations

The Microchip 16-bit Peripheral Library does a good job of implementing a useful framework of functions to control and access the SPI ports on the dsPIC DSC. Unlike the interfaces for the UART and the CAN bus that we'll look at shortly, the SPI is usually used to transfer small, often byte-size or word-size chunks of data. Because its transfer rate is so high, this means that we can essentially treat the transfers as in-line operations that are completed in real-time as the code that uses them executes. For instance, if we're reading two bytes of temperature data from a sensor connected via the SPI, often we can afford to issue the request and wait for the response since the data transfer will not significantly slow our operation. This is not the case when transferring large amounts of data through other communication ports (or even through the SPI); generally, we have to implement a buffered, interrupt-driven framework to deal with that situation. Fortunately, the Microchip 16-bit Peripheral Library already has all of the functionality we need to use the SPI port.

To use the SPI port, we first have to configure the associated dsPIC module, setting the data transfer rate, the clock polarity, the clock phase (these last two essentially set the operating mode), and the interrupt priority.

The Universal Asynchronous Receiver-Transmitter (UART)

While the SPI and the I²C interfaces are both synchronous interfaces, the Universal Asynchronous Receiver-Transmitter, or UART, not surprisingly uses an asynchronous link. The UART is probably the most ubiquitous means of communication, at least between computers, with the best example being the trusty RS-232 serial port that until recently was a standard part of any personal computer. It turns out that RS-232 serial ports are also widely used in industrial systems as well and were one of the earliest widely-used standard interfaces, along with their close cousins, the RS-422 and the RS-485 interfaces, which are essentially differential versions of the same basic interface. In fact, the "U" in UART reflects the "universal" nature of this form of communication.

UART-based devices can operate in either point-to-point topologies or in multinode topologies provided that additional driver circuitry is included that can disable each node's transmitter. The minimum number of signals required for UART communications is about as simple as it gets: a Receive Data (RxD) signal, a Transmit Data (TxD) signal, and a ground signal to which to reference the other two lines. In practice, the RxD signal at one node is connected to the TxD signal at the other node, and vice versa. As we've already discussed, asynchronous links embed the clock signal in the transmitted data, or they at least offer some way in which the receiver can synchronize its high-speed internal clock with the incoming data stream so it can extract the correct data bit values. For a UART-based interface, this synchronization is performed by insertion of a Start bit value at the beginning of each data byte value. The data itself follows next, starting with the least significant bit (LSB) and finishing with the most significant bit (MSB). Following the data is an optional parity bit, and a stop period that is 1, 1.5, or 2 bit periods long. Figure 4.6 illustrates a sample data transmission.

Sample UART Data Transfer

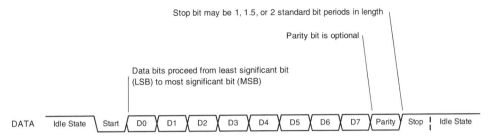

Stop bit may be 1, 1.5, or 2 standard bit periods in length

Parity bit is optional

Data bits proceed from least significant bit
(LSB) to most significant bit (MSB)

DATA | Idle State | Start | D0 | D1 | D2 | D3 | D4 | D5 | D6 | D7 | Parity | Stop | Idle State

---

Key Points

1. UART data transfers are characterized by
   A. Bit rate – number of bits per second
   B. Parity type – the manner in which the parity bit is generated
   C. Data size – number of data bits per transfer
   D. Number of stop bits

2. The bit rate, sometime erroneously referred to as the baud rate, determines the width of each data bit and is specified as the number of bits per second that the channel can transfer. Some common bit rates are 9600, 19.2K, 38.4K, 57.6K, and 115.2K bps. Although there is no requirement to use these transfer speeds (as long as both the receiver and the transmitter employ identical bit rates), failure to do so may prevent a system from communicating with other systems.

3. There are three types of parity: even, odd, and none. Even parity will set the Parity bit to 1 if the data itself (not including overhead bits such as the start or stop bits) contains an even number of bits that are set and clears the Parity bit to 0 if the data contains an odd number of bits that are set.

   Odd parity sets the Parity bit to 1 if the data contains an odd number of bits that are set and clears it to 0 if there are an even number of bits set.

   If a parity type of None is selected, the parity bit is not included in the data stream.

4. The data size may be either 8 bits or 9 bits, but the use of 9-bit data requires additional software overhead to generate and process.

5. The length of the Stop period, commonly referred to as the number of Stop bits, may be either 1, 1.5, or 2 standard bit periods. Although low-speed data transmissions in which individual data transfers are sent infrequently may tolerate a mismatch in the number of data bits configured in the transmitter and in the receiver, at higher rates the mismatch will cause data transfer errors.

Figure 4.6. Sample UART Data Transmission

Note that the idle state for the RxD line is high; the Start bit takes the line from its idle state to ground to signal the receiver that data will follow. This causes the UART to start its internal bit timing clock and the hardware state machine that identifies the time at which to sample the incoming data stream for each bit. Obviously, both ends of the link must be configured to run at the same transfer speed or the receiver's sampling circuitry will attempt to read the individual bit values at an incorrect time, producing invalid data at the receiver.

The optional parity bit can be used for basic transmission error detection, although many applications choose to ignore it. When it is employed, the value of the parity bit is set by the transmitter, and the receiver should verify that the parity bit value it detects is appropriate for the data sent. There are three parity configuration options:

1. None – no parity value is generated by the transmitter nor is that bit period used;

2. Even – when set, the number of "1" bits in the data itself is even, and when cleared, the number of "1" bits in the data is odd; and

3. Odd – when set, the number of "1" bits in the data itself is odd, and when cleared, the number of "1" bits in the data is even (essentially the opposite of even parity).

Because its usage imposes additional software overhead on the system, parity should only be employed when it will be verified at the receiving end. This can be particularly useful in high-noise environments in which there is a likelihood that data may be corrupted.

Finally, the stop bits at the end of the data byte are important because they ensure that the line goes to the idle state for a sufficient length of time for the receiver's synchronization hardware to reset. Typically, data links try to operate with as few stop bits as possible to increase throughput, but noisy environments or nodes that require a little extra time to process the received data may require additional stop bits.

UARTs work particularly well in links that need run longer distances than just between two PCBs, although they work well for that purpose, too. Typical maximum lengths for single-ended RS-232 links are around 15 meters, but differential RS-422 and RS-485 links are rated up to 4,000 meters. Speeds for UART interfaces are typically between 300 bps and 115 Kbps.

A Basic UART Interface Framework

The code for the UART interface is found in the file `CommIF.c`, with prototypes and associated definitions given in `CommIFDef.h`. Implemented as a buffered, interrupt-driven framework, the interface supports the transmission and reception of data from circular receive and transmit buffers maintained by the application. By using a buffered interface, the framework frees the application from the need to block on the transmission or reception of messages that are longer than the UART's internal 4-byte receive and transmit buffers, a feature that is mandatory in time-critical systems. In addition, the interface allows the designer to use either UART 1 or UART 2 when operating on a dsPIC DSC that supports two UART ports.

To use the interface, the application must first initialize the desired communication port by calling the routine CommInit() with parameters specifying the port to initialize and its associated operating parameters (speed, type of parity, and number of stop bits). Since the function initializes the global state variables used to control the interface as well as configuring the UART hardware, failure to invoke the routine before using the other UART interface functions will cause erratic behavior. The code for CommInit() is shown in Example 4.1.

Code Example 4.1. The CommInit() *Function*

```
/*************************************************************************
*    FUNCTION:         CommInit(Uint8 ui8Port, Uint16 ui16BaudRate,    *
*                            Uint16 ui16Parity, Uint16 ui16StopBits)   *
*                                                                       *
*    DESCRIPTION:      This function configures the system's communication *
*                      channel(s). It must be customized for the specific *
*                      communication modules and channel parameters used *
*                      for the particular application.                  *
*                                                                       *
*                      The function uses the global system communication *
*                      configuration parameters; if any of these parameters *
*                      are invalid, it uses the default channel parameters *
*                      to ensure that the communication channel is at least *
*                      operational.                                     *
*                                                                       *
*                      NOTE:   Although the dsPIC hardware allows 9-bit *
*                              data, this routine only supports 8-bit data *
*                              since that is more commonly used and is  *
*                              easier to manipulate.                    *
*                                                                       *
*    PARAMETERS:       ui8Port          - index of UART port to configure *
*                                        (MUST be either UART_1 or UART_2) *
*                      ui16BaudRate     - communication baud rate in bits/sec *
*                      ui16Parity       - type of parity to use (MUST be one *
*                                        of PARITY_*)                   *
*                      ui16StopBits     - number of stop bits (MUST be either *
*                                        STOP_BITS_1 or STOP_BITS_2)    *
*                                                                       *
*    RETURNS:          The function returns one of the following status *
```

```
*                   code values:                                        *
*                     ST_OK          - operation successful             *
*                     ST_INV_PARM    - invalid communication parameter  *
*                                      detected, default channel        *
*                                      parameters used                  *
*                     ST_COMM_INIT   - failed to initialize the requested *
*                                      communication channel(s)         *
*                                                                       *
*    REVISION: 0     v1.0                      DATE: 18 May 2006         *
*        Original release.                                              *
*************************************************************************/


Uint16
    CommInit(Uint8 ui8Port, Uint16 ui16BaudRate, Uint16 ui16Parity,
            Uint16 ui16StopBits)
    {
    //  Local Variables

    Uint16
        ui16BRGValue,       // Baud Rate Generator value
        ui16Mode,           // Configuration data for UxMODE register
        ui16Status;         // Configuration data for UxSTA register


    //  Log the UART that we're using
    //  for the communication port

    g_ui8CommPort = ui8Port;


    //  Turn off the specified UART module and
    //  disable the associated interrupt so we
    //  can complete the initialization without
    //  interruption

    if (ui8Port == UART_2)
        CloseUART2();                       // Close UART 2

    else
        CloseUART1();                       // Close UART 1


    //  Compute the Baud Rate Generator value based
    //  on the system instruction cycle frequency
    //  and the desired baud rate. From the 30F6014A
    //  data sheet, the BRG value is computed as
```

```
//     BRG = ((FCY / baud rate) / 16) - 1
//   where
//     FCY = instruction cycle frequency in Hz
//     baud rate = desired baud rate in bits/sec

ui16BRGValue = ((FCY / ui16BaudRate) / 16) - 1;

//  Setup the UxMODE register value to specify
//  whether the UART is enabled, how it will handle
//  the IDLE state, and other configuration data

ui16Mode = UART_EN            & // Enable the UART
           UART_IDLE_CON      & // Stop the UART when in the IDLE state
           UART_DIS_WAKE      & // Don't enable UART Wake on Start
           UART_DIS_LOOPBACK  & // Disable loopback mode
           UART_DIS_ABAUD;      // Don't use Auto-baud mode

//  Add the appropriate configuration information
//   for the number of data bits and parity to employ

switch (ui16Parity)
    {
    case PARITY_EVEN:
        ui16Mode &= UART_EVEN_PAR_8BIT; // Even parity, 8 data bits
        break;

    case PARITY_ODD:
        ui16Mode &= UART_ODD_PAR_8BIT;  // Odd parity, 8 data bits
        break;

    default:
        ui16Mode &= UART_NO_PAR_8BIT;   // No parity, 8 data bits
        break;
    }

//  Add the appropriate configuration information
//   for the number of stop bits to use

if (ui16StopBits == STOP_BITS_2)
    ui16Mode &= UART_2STOPBITS;     // Use 2 stop bits

else
    ui16Mode &= UART_1STOPBIT;      // Use 1 stop bit
```

```
//  Setup the UxSTA register value to specify
//  how the transmit and receive interrupts
//  will be generated and other processing
//  control configuration

ui16Status = UART_INT_TX         &    // Interrupt on Tx register
             UART_TX_PIN_NORMAL  &    // Use normal Tx pin state (not
                                      //   transmitting a break)
             UART_TX_DISABLE     &    // Disable the transmitter for now
             UART_INT_RX_CHAR    &    // Interrupt on character reception
             UART_ADR_DETECT_DIS &    // Don't use Address Detect mode
             UART_RX_OVERRUN_CLEAR;   // Clear the UART Rx Overrun flag


//  Initialize the global state variables
//  associated with the communication channel

g_ui16CommRxAppIndex = 0;       // Initialize the app-side Rx buffer index
g_ui16CommRxISRIndex = 0;       // Initialize the ISR-side Rx buffer index
memset(g_ui8CommRxData, 0, COMM_RX_BUFF_SZ);   // Initialize Rx buffer data

g_ui16CommTxAppIndex = 0;       // Initialize the app-side Tx buffer index
g_ui16CommTxISRIndex = 0;       // Initialize the ISR-side Tx buffer index
memset(g_ui8CommTxData, 0, COMM_TX_BUFF_SZ);   // Initialize Tx buffer data


//  Actually initialize the appropriate UART

if (ui8Port == UART_2)
    {
    ConfigIntUART2(UART_RX_INT_EN  & UART_RX_INT_PR6 &  // Enable Rx interrupt
                   UART_TX_INT_DIS & UART_TX_INT_PR3);  //   and disable Tx int

    OpenUART2(ui16Mode, ui16Status, ui16BRGValue); // Open UART 2
    }

else
    {
    ConfigIntUART1(UART_RX_INT_EN  & UART_RX_INT_PR6 &  // Enable Rx interrupt
                   UART_TX_INT_DIS & UART_TX_INT_PR3);  // and disable Tx int

    OpenUART1(ui16Mode, ui16Status, ui16BRGValue);      // Open UART 1
    }

return ui16Status;
}
```

Once the communication interface has been initialized by `CommInit()` and the interrupt level on the dsPIC DSC has been set low enough to allow interrupts from the serial port (interrupt level 6), the application can write and read data to and from the selected UART. To read data from the initialized serial port, the application first checks to see whether there is data pending in the global Receive Data queue by calling `CommGetRxPendingCount()`, which returns the number of unread received data bytes pending in the Receive Data queue. If this value is nonzero, indicating that there is data available to read, the application then reads the next pending character from the queue by calling the function `CommGetRxChar()`. Figure 4.7 shows the flow chart illustrating this sequence.

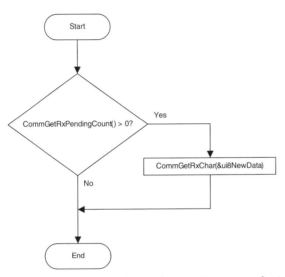

Figure 4.7. Flow Chart for Reading UART Data in the Application

Assuming that the `CommGetRxChar()` routine returns a status code value of ST_OK, the application can safely use the data read from the UART. If, however, the status code value is not ST_OK, the application should not use the data in ui8NewData, since it will not be valid.

Transmitting data is also a simple matter. Assuming that the communication interface has been initialized by calling CommInit(), the application uses Com-mPutChar() and `CommPutBuff()` to transmit either a single byte of data or multiple bytes of data, respectively. A quick look at the code shows that Com-mPutBuff() simply calls `CommPutChar()` repeatedly to load the entire source data contents into the circular transmit buffer, but an examination of the code for `CommPutChar()` (shown in Code Example 4.2) is instructive.

Code Example 4.2. The CommPutChar() *Function*

```
/************************************************************************
*     FUNCTION:       CommPutChar(Uint8 ui8Data)                        *
*                                                                       *
*     DESCRIPTION:    This function adds a character to the global      *
*                     Transmit Data buffer for subsequent transmission  *
*                     by the UART's Transmit Data interrupt handler.    *
*                     If the buffer is full, the character is not added *
*                     to the buffer, and the function returns an error  *
*                     status.                                           *
*                                                                       *
*     PARAMETERS:     ui8Data - data to be added to the transmit data   *
*                               buffer for transmission by the USART    *
*                                                                       *
*     RETURNS:        The function returns one of the following status  *
*                     code values:                                      *
*                     ST_OK          - operation successful             *
*                     ST_BUFFER_FULL - transmit buffer full, data not   *
*                                      added for transmission           *
*                                                                       *
*   REVISION: 0    v1.0                    DATE: 18 May 2006            *
*       Original release.                                               *
************************************************************************/

Uint16
    CommPutChar(Uint8 ui8Data)
    {
    //  Local Variables

    Uint16
        ui16AppIndex,       // Temporary buffer for processing the
                            //    application index for the transmit
                            //    data buffer
        ui16DataCount,      // Number of characters currently in
                            //    the transmit data buffer
        ui16Status;         // Function execution status

    //  Check whether there is room in the
    //  transmit data buffer for the new data

    ui16Status    = ST_BUFFER_FULL; // Assume the transmit buffer is full
    ui16DataCount = CommGetTxBuffCount();   // Get the number of characters already
                                //    in the transmit data buffer
```

```
if (ui16DataCount < (COMM_TX_BUFF_SZ - 1))
    {
    //  We do have room in the buffer,
    //  so add the new character to it

    g_ui8CommTxData[g_ui16CommTxAppIndex] = ui8Data;  // Add the new
                                                      //    character


    //  Modify the application-side index to point
    //  to the next character in the transmit buffer.
    //  Use a local variable to do the manipulations
    //  and then transfer value to the global state
    //  variable in one operation to prevent coherency
    //  problems with the ISR.

    ui16AppIndex = g_ui16CommTxAppIndex;    // Get the current
                                            //   application index
    ui16AppIndex++;                         // Move to the next slot
    ui16AppIndex &= (COMM_TX_BUFF_SZ - 1);  // Perform a rapid modulo-
                                            //   COMM_TX_BUFF_SZ
                                            //   calculation
                                            //   (assumes that COMM_TX_
                                            //   BUFF_SZ
                                            //   is a power of 2)


    //  Update the state variable that controls
    //  access to the global Transmit Data queue

    g_ui16CommTxAppIndex = ui16AppIndex;

    //  Make sure that the hardware can and will send
    //  the data. Since the data is actually loaded
    //  into the UART's Transmit Buffer in the ISR,
    //  we need to make sure that the ISR is called
    //  if the Transmit Buffer has room. If there is
    //  no room in the UART's Transmit Buffer, we don't
    //  need to generate an interrupt at this time (the
    //  new data has already been added to the global
    //  Transmit Data queue and will be loaded into the
    //  UART's Transmit Buffer at a later interrupt.)

    CommEnableTransmitter();    // Make sure the transmitter is enabled -
                                //   must occur FIRST
```

```
      if (CommTxBuffAvail())

          CommForceTxInt();        // UART's Transmit Buffer has room so
                                   //   force a Transmit interrupt. This
                                   //   must be performed BEFORE making sure
                                   //   that the UART's Transmit interrupt
                                   //   is enabled to avoid a race condition
                                   //   with the Transmit Data ISR.


          CommEnableTxInt();       // Finally, ensure that the UART's
                                        Transmit
                                   //   interrupt is enabled


      //  Flag that the operation was successful

      ui16Status = ST_OK;
      }


  return ui16Status;
  }
```

The Controller Area Network (CAN)

The Controller Area Network is the probably the most sophisticated of the serial interfaces offered on the dsPIC DSC. It incorporates a very advanced internal hardware controller that supports moderate speed (up to 1 Mbps) data transfers with built-in hardware error detection, a sophisticated message prioritization scheme, and the ability to set filters that allow only messages of interest to be received, all with very little processor overhead. Widely used in the automotive and industrial-processing world, the CAN architecture offers a robust way to link together multiple nodes on a single network.

With all of these positives, why would anyone *not* use the CAN interface? There are two main reasons: complexity and cost. One big advantage of CAN is that it's highly configurable, but one big disadvantage is that CAN is so highly configurable. Because it's so flexible, a CAN topology can be used in a wide variety of applications using basically the same hardware. Unfortunately, that flexibility must be configured fairly precisely or the channel will be either unreliable or completely unusable, and debugging problems with the channel can be both time-consuming and frustrating.

Basic CAN Architecture

Developed by Bosch in the early 1980s, the CAN architecture is pretty simple. Although the CAN standard itself is intentionally media-neutral,[2] one of the most common implementations uses a single differential serial bus running at 1 Mbps[3] or less to connect two or more nodes together. Along with the associated ground signal, a reliable interface can consist of only three wires!

The CAN's communication protocol is a member of the CSMA/CD family, a cryptic acronym that stands for Carrier Sense Multiple Access/Collision Detection. Although the family name is long, the concepts behind it are easy. In a *carrier sense* system, all nodes have to monitor the network for a period of inactivity before they can attempt to send a message. Once this inactive period has elapsed, however, any of the nodes in the network can transmit data, hence the term *multiple access*. As one would expect, there will be times when two or more nodes try to send data at the same time, a condition known as *collision*, so the network has to have some way to perform *collision detection*. Individual members of the CSMA/CD family handle these tasks differently, but all members of a given type (such as CAN) do so in the same manner.

Of these tasks (carrier sensing and collision detection), the more difficult by far is collision detection. The CAN designers came up with an ingenious solution to this problem, one that allows the system designer to prioritize message traffic so that more important messages are always able to gain access to the bus ahead of less important messages (in much the same way that interrupts are prioritized by the dsPIC DSC). Not only does the CAN allow message prioritization, its network *arbitration*[4] scheme is *nondestructive*[5] to the higher priority message and ensures that the higher priority message experiences no transmission delay. Since message arbitration is so important, we'll look at that in detail after we first get some more background information under our belt.

Another of the CAN's key features is its built-in error-detection circuitry that flags problems with the bus and that will gradually remove an individual network node from the bus should the node generate too many errors. Although the protocol does not support error correction, its error-detection feature helps avoid the serious problem of a single erroring node bringing down the entire network. Unfortunately, because errors can accumulate quickly when there is a problem, tracking down the source of the problem can be difficult since it may go away once the node stops trying to transmit.

If all of this functionality sounds as though it imposes a severe load on the processor, you can relax; because of its complexity, the vast majority of the CAN interface is contained in two hardware components: a CAN controller state machine that handles all of the arbitration and error detection and a CAN bus driver that drives and monitors the CAN bus physical medium. In most systems, these two hardware components are housed in individual integrated circuit (IC) packages.[6] This is the case with the dsPIC DSC, which contains either one or two CAN controller modules on-chip (depending upon the dsPIC device) and which requires an external CAN driver chip to connect to the bus. Once the CAN interface circuitry has been configured, it simply presents fully formed data messages and status bits to the receiver and transmits complete data messages to other nodes. Since all error detection and handling is performed in hardware, the processor overhead associated with the CAN interface is minimal.

One last high-level consideration is just how far one can run a CAN bus, and the answer is that the maximum bus length depends upon the data rate that the bus must support. Table 4.1 shows the recommended maximum bus lengths for a variety of bit rates.[7]

Bit Rate (Kbps)	Bus Length (m)
1,000	30
500	100
250	250
125	500
62.5	1,000

Table 4.1. Recommended Maximum CAN Bus Lengths

As the table clearly demonstrates, the maximum bus length drops off rapidly with increasing data rates, but even at 1 Mbps (1,000 Kbps), the maximum bus length is reasonably robust.

CAN Data Formats

According to the CAN 2.0 specification,[8] data sent over the CAN bus is in one of four basic data formats, called frames:

1. the Data frame, which transmits data from one node to all other nodes on the bus,

2. the Remote Transfer frame, which requests data from another node on the bus,

3. the Error frame, which reports that a communication error has been detected, and

4. the Overload frame, which reports that the transmitting node is busy processing a previous message and cannot accept more data at this time.

In our example, we will be interested primarily in the most commonly used format, the Data Frame, which comes in two flavors: the standard frame and the extended frame. The two data frame formats, illustrated in Figures 4.8a and 4.8b, are essentially identical, with the only real difference being the shorter arbitration ID of the standard frame. All data frame formats have the following basic elements:

1. an Arbitration ID field whose size varies with the frame type,

2. a 6-bit Control Field,

3. a Data Field of 0 to 8 bytes in length,

4. a 2-byte CRC Field,

5. a 2-bit Acknowledge Field, and

6. a 1-bit End of Frame marker.

Of these fields, the user has control over the arbitration ID, the Control, and the Data fields, while the CAN controller hardware automatically generates and validates the CRC Field, the Acknowledge Field, and the End of Frame marker. Let's delve a little deeper into the fields before examining the CAN arbitration technique.

In a standard data frame, the Arbitration ID field consists of an 11-bit identifier and a 1-bit Remote Transmission Request (RTR) flag. The extended data frame format is slightly different, but it is designed so that if there is a collision between a standard frame and an extended frame, the standard frame has priority. For an extended frame, the identifier is 29 bits, with the 11 most significant bits being transmitted after the Start of Frame, followed by a 1-bit Substitute Remote Request (SRR) flag, a 1-bit Identifier Extension (IDE) flag, and then the remaining 18 bits of the identifier, with the 1-bit RTR flag completing the field.

There are also slight differences in the Control Field layout for the two data frame formats, although it is 6 bits wide in both cases. In the standard frame, the leading bit of the Control Field is the IDE flag, which is followed by a single reserve bit denoted as "r0" in the CAN specification. The final four bits of the field comprise the Data Length Code (DLC), which specifies the number of data bytes that will follow in the message. Although the DLC is four bits wide, it can only assume a value of 0 to 8, since the protocol supports a maximum of 8 data bytes per message.

Standard (11-bit ID) CAN Data Frame Format

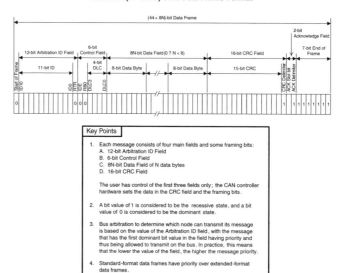

Figure 4.8a. Standard CAN Data Frame Format

Extended (29-bit ID) CAN Data Frame Format

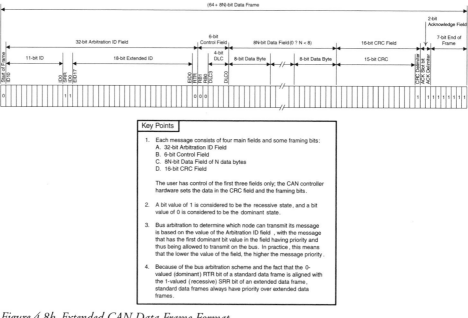

Figure 4.8b. Extended CAN Data Frame Format

Because the extended frame includes the IDE flag as part of the Arbitration ID field, it has two reserved bits in the Control Field, "r1" and "r0". The DLC is the same as in the standard data frame and labors under the same restrictions.

The Cyclic Redundancy Code (CRC) field is not of much interest to us as designers, since it is handled exclusively in hardware and is therefore transparent to the programmer. For sake of completeness, let us note that the CRC itself is a 15-bit value, and the CRC field is composed of the CRC value and a 1-bit CRC delimiter bit.

The final field in a CAN message is the 2-bit Acknowledgement (ACK) Field, which consists of a leading ACK Slot bit that is set to the recessive state (which will be defined in the next paragraph) by the transmitting node and then set to the dominant state by all nodes that receive the message successfully, whether they actually use the message or not. The final bit in the ACK Field (and the message) is the ACK delimiter bit, which simply returns the bus to the recessive state to signal that the transmission is complete.

Although we won't use Remote Transfer frames, Error frames, or Overload frames in our example, the dsPIC DSC's CAN interface is fully capable of handling these. Remote Transfer frames are used to request the automatic transmission of data from a node (the data having been already loaded into the CAN module in anticipation of the request), and Error frames are generated by a node when it detects an error condition on the bus. Because Error frames intentionally violate the timing parameters of the CAN bus, they cause all of the nodes that were transmitting data to stop, reset the transmission, and start their transmissions again.

Bus Arbitration

As we've already noted, since data transfers are asynchronous, some sort of access arbitration is required to determine which node may transmit if two attempt to send data simultaneously. The CAN designers came up with an ingenious solution to this problem, creating a nondestructive arbitration scheme that uses the value of the arbitration IDs of the colliding messages to decide which node has priority. To understand how this scheme works, we first need to learn two terms that apply to CAN-based systems. Data on the CAN bus is said to be in either a dominant state (a logical 0) or a recessive state (logical 1). When two bits of different state are transmitted at the same time, the dominant state "wins,"—i.e., that is the resulting state on the bus.

The CAN uses this fact for its transmission access arbitration. Whenever two or more nodes try to transmit a message simultaneously, the dominant bit state is the one that is present on the bus. As each node transmits data onto the bus one bit at a time, it checks to see whether the data on the bus reflects the state of the most recently transmitted bit. If a transmitting node sends a recessive bit but detects that the bus is in the dominant state, the node knows that there is another node that is also transmitting, and the node whose data was recessive knows to get off of the line. The recessive node immediately disables its transmitter and waits until the end of the current transmission before attempting to transmit its own data again.

By handling the arbitration in this manner, the CAN assures both that there is a structured approach to transmission access and that collisions don't result in lost data that forces all nodes to retransmit their messages. Since the dominant state is 0, designers of CAN-based systems select arbitration IDs such that the most important messages have low ID values and thus the highest priorities. For instance, by choosing arbitration ID values of 000H–01FH for alarm conditions and ID values of 020H–7FFH for normal operating messages, the designer ensures that alarm messages always have priority over normal operating messages. The example shown in Figure 4.9, in which an alarm message with an arbitration ID value of 010H is sent at the same time as a normal operating message with an arbitration ID value of 040H, illustrates this. In addition, the scheme allows both standard and extended data frames to reside on the bus, with the standard frame messages having priority over the extended data frames.

CAN Message Bus Arbitration

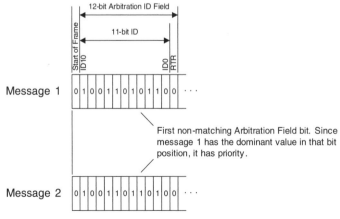

Figure 4.9. Example Arbitration of Two Simultaneous Messages

Acceptance Filters

One optional aspect of the CAN protocol that all CAN controllers implement is *message filtering*, which allows the controller to accept only messages whose Arbitration ID fields match a programmable bit-mapped filter value. In this case, when we refer to a filter, we're not talking about a digital filter that processes the digitized signal; rather, we're referring to the process by which only a limited group of messages that meet certain criteria are selected for processing by the CAN controller. Note that, even when the controller chooses to ignore the message, it always responds with the ACK Slot bit set appropriately. Filtering is a midlevel technique by which we can reduce the overhead on the processor by limiting the types of messages we choose to handle, while the acknowledgement process is a low-level requirement for ensuring the accurate delivery of the network traffic to all nodes.

Filtering the CAN messages consists of two steps, both of which are configurable by the designer but which are executed by the CAN controller hardware. First, we need to set the acceptance filter values (the dsPIC DSC supports up to six different filters), which are logically ANDed with the Arbitration ID of each received message on a per-bit basis. The resulting value is then compared to an acceptance mask on a per-bit basis, and if the result of applying the filter matches the acceptance mask, the incoming message is added to the CAN receive buffer (assuming there's room in the buffer).

This can be a point of significant confusion for new (and sometimes more experienced) CAN designers, so an example is appropriate. Let's assume that we want to accept any standard CAN data frame whose Arbitration ID field is in the range of 300H to 3FFH. In that case, the acceptance filter is simply F00H and the acceptance mask is also 300H, since ANDing the acceptance filter with the 12-bit Arbitration ID field of any received message will make the lower byte of the Arbitration ID field a don't care condition (since the entire lower byte will be ANDed with 0), and the filter will pass through the upper nibble. Only Arbitration ID fields whose upper nibbles are equal to 3 will match with the acceptance mask and thus be accepted.

Basic CAN Interface Framework

The Microchip 16-bit peripheral libraries do an excellent job of providing the tools a programmer needs to interface to the dsPIC DSC's CAN controller(s), but as in the case of the UART interface, they don't directly implement interrupt-driven buffered message (not character, since the CAN is a message-based protocol) I/O. Fortunately, we can simply tweak the code that we developed for the UART, and we have a similar framework for the CAN interface. This is an excellent example of the value of writing code that is somewhat generic; since the same basic principles apply to both the UART and the CAN interfaces, we're able to reuse the same basic code in the CAN that we developed for the UART (after making the requisite modifications to account for the differences in the two modules, of course).

4.3   High-level Protocols

So far we've looked at the basic communication channels essentially as I/O mechanisms that transfer data between the dsPIC DSC and other devices, and our focus has been on the bit- or the byte-level. Even when we talk about the CAN protocol, what we're really examining is the low-level hardware interface and the set-up information required to properly configure it. All of this is necessary, but it's not sufficient; we also need to define what the data that is transferred actually means, and thus we turn now to the issue of *high-level protocols*, or *HLPs*.

In earlier chapters, we've discussed the need to turn data into information, and the structure imposed by a high-level protocol allows us to do precisely that. The situation is analogous to sending a letter from one person to another. In order for the letter to reach its destination, we have to specify certain information (the recipient's address) and we may add optional information (such as the recipient's name) depending on the circumstances. Since there may be a problem with the delivery, it's also a good idea to include information that can be used to report or recover from the problem (for example, the sender's name and address). This is exactly what we have to do with the data that we send between nodes in the system, whether those transmissions are between individual devices like the dsPIC DSC and an external DAC or whether they are between the dsPIC DSC and remote systems. By imposing a structure on the data that we transfer, we allow that data to be understood clearly and processed efficiently.

A number of popular standard protocols are available for serial data and, depending upon the type of equipment with which the sensor must communicate, the designer may have no alternative but to implement a particular protocol for the application. If the protocol is widely used, this approach offers the likelihood (or at least the possibility) of having available third-party software implementations that can be dropped into the application and/or tools that can be used to test the resulting code. The downside occurs when the application only requires a subset of functionality to accomplish its task, because then the full functionality of the standard protocol wastes precious processor resources. Sometimes this is the only option—for example, when one is communicating with another chip via the SPI. Usually, individual devices have a specific protocol that must be used to communicate with them, leaving the designer with little or no flexibility.

An alternative approach, and the one employed in this example for communicating between a host system and the dsPIC DSC, is to create a proprietary protocol that is tailored to the specific application. Designed appropriately and properly implemented, a proprietary protocol can offer the required functionality without wasting resources on features that will never be used by the application; however, it's always a good idea for the designer to ensure that the application actually requires (and the development team can afford) the time and resources required to develop a proprietary solution.[9]

If, after looking at the situation objectively, the designer decides that a proprietary protocol is necessary, one should always try to make it as *lightweight*, as *reliable*, and as *extensible* as possible. By light-weight, we mean that the protocol should add as little data and processing overhead as possible while still accomplishing the task in a reliable manner. Although the concept of reliability is self-explanatory, we also include error detection as a desirable goal as well. Finally, a protocol's extensibility refers to the ability to add new message types easily and in a manner that has minimal impact on existing code.

The protocol used in the examples demonstrates one such implementation (there are others) that balances the need for low processing overhead, error detection, and extensibility. The reader should understand that this is not the only way, nor perhaps even the best way, for a specific application; it does, however, work well in a variety of circumstances. In actuality, there are two protocols at work in these examples: one protocol that encapsulates the general message and a second that is message-type specific. Let's start by examining the general message protocol first.

General Message Protocol

The general message protocol is a simple command/response protocol that has one primary goal: to deliver variable-length command messages to the target system and to process the returned variable-length response message. The simple protocol illustrated in Figures 4.11 and 4.12 handles this goal nicely.

Start Token (1 byte)	Command ID (1 byte)	Data Length (1 byte)	Data (N bytes)	Checksum (1 byte)
0x01 (SOH)	-	N	-	Calculated

Figure 4.10. General Message Protocol Command Format

For a command,

Start Token	start of command token (ASCII Start of Header, 0x01)
Command ID	the ID of the command being sent (MSB set to denote a command)
Data Length	length of the data associated with the command (N bytes, where $0 \leq N \leq 255$
Data	N bytes of command data (not used if $N == 0$)
Checksum	checksum computed from the Command ID, Data Length, and all Data bytes

The format for a command response message is shown in Figure 4.12.

Start Token *(1 byte)*	*Response ID* *(1 byte)*	*Status* *(1 byte)*	*Data Length* *(1 byte)*	*Data* *(M bytes)*	*Checksum* *(1 byte)*
0x01 (SOH)	-	-	*M*	-	*Calculated*

Figure 4.11. General Message Protocol Response Format

where,

Start Token	start of command token (ASCII Start of Header, 0x01)
Response ID	the ID of the response being returned (equal to the corresponding Command ID value but with the MSB cleared to denote a response)
Data Length	length of the data associated with the response (M bytes, where $0 \leq M \leq 255$, and N and M may not be the same)
Data	M bytes of response data (not used if M equals 0)
Checksum	checksum computed from the Response ID, Data Length, and all Data bytes

Since one can (and should) assume that errors will occur at some point in the communication process, the protocol employs a start character to signal the start of a new transmission and a checksum[10] value to verify that the data received was, in fact, the same as the data transmitted. This not only offers a degree of error detection but also the ability to resynchronize the data between the transmitter and the receiver should an error occur, since the receiver can be looking for the start character. Although there are many different ways to compute a checksum, the method used in these examples is to add all of the data bytes between (but not including) the start character and the transmitted checksum value itself, and then to perform a bit-wise inversion of the summed value. The advantage of inverting the sum rather than simply using the sum itself is that a long string of 0s produces a checksum value of FFH rather than a checksum of 00H, which forces a differentiation between the checksum and the data. If there were no differentiation, a signal stuck at 0 would not be detected since both the data and the checksum values would be 0 as well.

To further identify the commands, the command ID values all have the most significant bit (MSB) set to 1, and the corresponding responses have the response ID identical to the command ID but with the MSB cleared. This makes debugging the communications just a bit easier for the designer, since she can pick out the command and response messages easily on a data analyzer. It also provides another way to validate a command message when it is received.

As we've discussed previously, it's important to implement the protocol parser as a state machine so that we never get into a state from which we cannot recover. This also keeps us from making too many foolish assumptions, such as "the message will always be a certain length." While it may be true that the message is *supposed* to be a certain length, it's also equally true that there will be times that for some reason the received data is *not* the anticipated length, and we can't allow that to then stop future communications. Figure 4.13 shows a flow chart of the protocol parsing state machine that we'll use.

Command-specific Protocols

Each command has its own set of data that it sends to the receiver, and the response that the receiver reports after processing the command message is command-dependent as well. Although the specific commands and responses are somewhat application-dependent, those common to all three applications are shown here to give the reader examples of such an implementation. Note that the command-specific protocol does not perform any error checking. It is the responsibility of the general message protocol to ensure that accurate data is delivered to the receiver; once that data has arrived, it is assumed to be accurate. Functions that use the parsed data should, of course, verify that the data is valid in the sense of being appropriate for the specific task. As an example, if the host sent a temperature setpoint of 1500 °F, but the maximum allowed setpoint was only 1000 °F, the received value of 1500 °F would be accurate (i.e., would be the value actually transmitted by the sender) but invalid (not an allowed value for this parameter). Table 4.2 lists the command message data formats for those messages that are common across of the example applications, while Table 4.3 lists the corresponding response message formats.

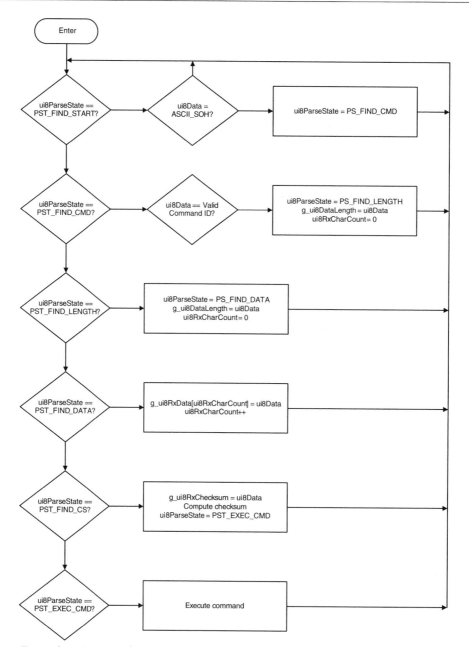

Figure 4.12. State Machine to Process Protocol

Command ID	Data Length	Parameter Values
0x80	0	Report release information

0x81	5	Log lower calibration measurement
		    Parm 0 – 8-bit 0-based sensor index
		    Parm 1 – 32-bit defined lower calibration value

0x82	5	Log upper calibration measurement
		    Parm 0 – 8-bit 0-based sensor index
		    Parm 1 – 32-bit defined upper calibration value

0x83	1	Compute calibration gain and offset Parm 0 – 8-bit 0-based sensor index
0x84	1	Report calibration gain Parm 0 – 8-bit 0-based sensor index
0x85	1	Report calibration offset Parm 0 – 8-bit 0-based sensor index

0x86	6	Configure lower limit alarm
		    Parm 0 – 8-bt 0-based sensor index
		    Parm 1 – 32-bit alarm limit
		    Parm 2 – 8-bit alarm enable flag

0x87	6	Configure upper limit alarm
		    Parm 0 – 8-bt 0-based sensor index
		    Parm 1 – 32-bit alarm limit
		    Parm 2 – 8-bit alarm enable flag

0x88	1	Reset alarms Parm 0 – 8-bit 0-based sensor index
0x89	1	Report alarm states Parm 0 – 8-bit 0-based sensor index
0x8A	1	Report lower alarm limit Parm 0 – 8-bit 0-based sensor index
0x8B	1	Report upper alarm limit Parm 0 – 8-bit 0-based sensor index
0x8C	1	Report sensor value Parm 0 – 8-bit 0-based sensor index
0x8D	1	Set digital potentiometer value Parm 0 – 8-bit digital potentiometer value

Table 4.2. Command Message Data Formats

Response ID	*Data Length*	*Parameter Values*
0x00	6	Report release information Parm 0 – 8-bit release date month Parm 1 – 8-bit release date day Parm 2 – 8-bit upper two digits of release date year Parm 3 – 8-bit lower two digits of release date year Parm 4 – 8-bit major release version ID Parm 5 – 8-bit minor release version ID
0x01	3	Log lower calibration measurement Parm 0 – 8-bit 0-based sensor index Parm 1 – MSB of 2-byte fractional value measured at calibration point Parm 2 – LSB of 2-byte fractional value measured at calibration point
0x02	3	Log upper calibration measurement Parm 0 – 8-bit 0-based sensor index Parm 1 – MSB of 2-byte fractional value measured at calibration point Parm 2 – LSB of 2-byte fractional value measured at calibration point
0x03	3	Compute calibration gain and offset Parm 0 – 8-bit 0-based sensor index
0x04	5	Report calibration gain Parm 0 – 8-bit 0-based sensor index Parm 1 – MSB of MSW of 4-byte IEEE-754 floating point gain value Parm 2 – LSB of MSW of 4-byte IEEE-754 floating point gain value Parm 3 – MSB of LSW of 4-byte IEEE-754 floating point gain value Parm 4 – LSB of LSW of 4-byte IEEE-754 floating point gain value
0x05	5	Report calibration offset Parm 0 – 8-bit 0-based sensor index Parm 1 – MSB of MSW of 4-byte IEEE-754 floating point offset value Parm 2 – LSB of MSW of 4-byte IEEE-754 floating point offset value Parm 3 – MSB of LSW of 4-byte IEEE-754 floating point offset value Parm 4 – LSB of LSW of 4-byte IEEE-754 floating point offset value
0x06	3	Configure lower limit alarm Parm 0 – 8-bt 0-based sensor index Parm 1 – 32-bit alarm limit Parm 2 – 8-bit alarm enable flag

Response ID	Data Length	Parameter Values
0x07	6	Configure upper limit alarm Parm 0 – 8-bt 0-based sensor index Parm 1 – MSB of MSW of 4-byte IEEE-754 floating point alarm limit value Parm 2 – LSB of MSW of 4-byte IEEE-754 floating point alarm limit value Parm 3 – MSB of LSW of 4-byte IEEE-754 floating point alarm limit value Parm 4 – LSB of LSW of 4-byte IEEE-754 floating point alarm limit value Parm 5 – 8-bit alarm enable flag
0x08	1	Reset alarms Parm 0 – 8-bit 0-based sensor index
0x09	3	Report alarm states Parm 0 – 8-bit 0-based sensor index Parm 1 – MSB of 2-byte bit-mapped alarm flags Parm 2 – LSB of 2-byte bit-mapped alarm flags
0x0A	6	Report lower alarm limit Parm 0 – 8-bit 0-based sensor index Parm 1 – MSB of MSW of 4-byte IEEE-754 floating point alarm limit value Parm 2 – LSB of MSW of 4-byte IEEE-754 floating point alarm limit value Parm 3 – MSB of LSW of 4-byte IEEE-754 floating point alarm limit value Parm 4 – LSB of LSW of 4-byte IEEE-754 floating point alarm limit value Parm 5 – 8-bit alarm enable flag
0x0B	6	Report upper alarm limit Parm 0 – 8-bt 0-based sensor index Parm 1 – MSB of MSW of 4-byte IEEE-754 floating point alarm limit value Parm 2 – LSB of MSW of 4-byte IEEE-754 floating point alarm limit value Parm 3 – MSB of LSW of 4-byte IEEE-754 floating point alarm limit value Parm 4 – LSB of LSW of 4-byte IEEE-754 floating point alarm limit value Parm 5 – 8-bit alarm enable flag
0x0C	5	Report sensor value Parm 0 – 8-bit 0-based sensor index Parm 1 – MSB of MSW of 4-byte IEEE-754 floating point sensor value Parm 2 – LSB of MSW of 4-byte IEEE-754 floating point sensor value Parm 3 – MSB of LSW of 4-byte IEEE-754 floating point sensor value Parm 4 – LSB of LSW of 4-byte IEEE-754 floating point sensor value
0x0D	1	Set digital potentiometer value Parm 0 – 8-bit digital potentiometer value

Table 4.3. Response Message Data Formats

4.4 Summary

We've covered a lot of ground in this chapter, looking at three different communication interfaces that are widely used to connect the dsPIC DSC to peripheral devices, to networks of sensors and controllers, and to remote systems. Although it may not seem so to the reader, we have of necessity touched rather lightly on the CAN interface (the *Family Reference* devotes 73 pages to this module alone), and we've really only hinted at the communication possibilities available to the designer given the profusion of standard and proprietary protocols to which the sensor may be asked to connect. Nonetheless, with these tools, the designer should have a solid foundation upon which to build and extend the communication capabilities of dsPIC-based systems.

Endnotes

1. In this case, *topology* is simply the technical term for the arrangement of nodes in a network.

2. *Media-neutral* just means that the protocol does not specify the physical medium required to implement the protocol. This was intentionally left out of the specification so that the protocol can operate over a variety of physical media (so long as the media supports the ability to have a dominant and a recessive bit state).

3. 1 Mbps = 1,000,000 bits per second

4. In this case, arbitration is the process by which one of two or more nodes that are competing for access to the network is allowed to transmit data. Interrupt arbitration is the process by which the dsPIC DSC's interrupt controller determines which interrupt condition to service.

5. *Nondestructive* arbitration means that the message that ultimately is transmitted on the bus is left intact. Destructive arbitration would determine which message should be allowed onto the bus, but it would corrupt the message, meaning that the node that is allowed to transmit would have to resend the message from the beginning, which adds to the overall transmission time and reduces the resulting available bandwidth.

6. Integrated circuits, or *ICs* as they're more commonly called, are the silicon chips that contain much of the electronic circuitry in a system.

7. This table is taken from Microchip Application Note 713 – *Controller Area Network (CAN) Basics*, which is available on the Microchip website (document DS00713A).

8. *CAN Specification 2.0*, Robert Bosch GmbH, 1991.

9. Many engineers (and companies) have a very bad case of NIHS (Not Invented Here Syndrome), a crippling affliction that causes its victims to reject design solutions simply because they were not created by the engineers themselves or by their companies. Although the topic is often addressed humorously, the consequences are anything but funny: significant development delays, missed market windows, and inferior products. These consequences cost companies a tremendous amount of added expenses and lost profits, so the decision to develop a proprietary solution should be one based on objective facts, not emotion.

10. A *checksum* is a common technique for verifying that data received at the destination is the data that was sent. The term comes from the fact that checksums are usually computed by some variation of adding together the individual data bytes in a transmission; the computed sum is then checked against the transmitted value.

A Basic Toolkit for the dsPIC DSC

*Every day you may make progress. Every step may be fruitful.
Yet there will stretch out before you an ever-lengthening, ever-
ascending, ever-improving path.*

—*Sir Winston Churchill*

With the basic understanding of the dsPIC DSC's functional modules that we developed in the previous two chapters, we're now ready to create a basic toolkit of software modules that we can use to create a flexible generic framework for implementing intelligent sensor applications. For our application development, we'll use a number of software tools and hardware development platforms available for free or at low cost from Microchip. Although there are certainly other good development tools available from third party-companies, these tools pass three important tests: they're readily available, they're inexpensive, and they work.

5.1 The Application Test Bed

One of the most frustrating aspects of real-world product development is attempting to bring up new, untested software on new, untested hardware. Errors in either the code or the hardware can bring everything to a grinding halt, often with no solid clues as to the source of the problem. To alleviate this issue, we'll develop our applications using the Microchip dsPICDEM 1.1 General Purpose Development Board (GPDB),[1] which conveniently uses the dsPIC30F6014A chip. Of course, in most circumstances the final product would use custom hardware that addresses the cost, size, power, and feature requirements for the specific application, but early development on a standardized, known-good hardware platform allows coding to begin prior to having the final system hardware, and it eliminates one major source of error (the hardware) when debugging the application.

We need some way to get our code onto the GPDB for testing, and any significant software development also requires some way to examine the operation of the application as it's actually running. Using the Microchip ICD 2 in-circuit debugger and Microchip's MPLAB IDE (integrated development environment), we can accomplish both those tasks. The IDE offers a convenient PC-based software development environment that includes an editor, simulator, and assembler. Although there are a number of good C compilers available for the dsPIC DSC, the example code is written specifically for the Microchip C30 compiler v2.02, a student edition of which can be downloaded free of charge from the Microchip website.[2] One advantage of the C30 compiler is that it integrates directly with the MPLAB IDE, though there are certainly other good compilers that have this ability, too. The ICD 2 must be purchased,[3] but the MPLAB IDE is available as a free download from Microchip.[4]

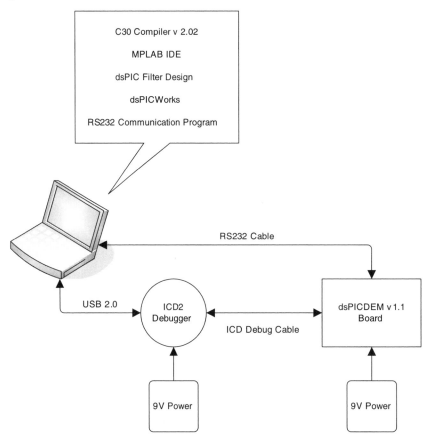

Figure 5.1. dsPIC Test Bed Block Diagram

5.2   Overview of the Firmware Framework

Now that we've set up a testbed, it's time to create the firmware framework that will serve as the foundation for the applications developed in the remainder of the book. In developing the code, we will implement the following design principles to ensure that the code is robust and that it anticipates and accounts for both normal and anomalous operation:

1. The code will use software state machines that are error-detecting and self-recovering in the event of problems. These software state machines will be as independent as possible and will employ simple, well-defined and well-behaved interfaces to other software components in the application.

2. To the extent possible, the firmware uses standard code libraries available from Microchip to trim the time required to code the application and to debug it. This also eases the maintenance issue since we'll be using previously tested code.[5]

3. The code is written in layers, particularly that for the communication interfaces and any that accesses the low-level hardware modules. This helps prevent changes in one section of the code from breaking other, unrelated sections of the code and gives the designer maximum flexibility in coding specific areas of the application.

4. All of the code is well-documented internally, with extensive, meaningful comments and standardized naming conventions to help developers who have to maintain the code in the future. Although the code fragments that are presented in the text have many of the comments stripped out for space reasons, the actual code included on the CD-ROM that accompanies the book has the comments restored.

Of these, the first and last design principles are extremely important, yet they are often ignored as being too time-consuming or unnecessary. By implementing self-correcting software state machines to handle the processing, we ensure (at least to the best of our ability) that the code will never get into a processing state from which it cannot recover. Doing so requires imagination and discipline since the designer must anticipate all possible anomalous conditions and construct techniques to both detect them and to recover from them should the conditions occur, but it pays tremendous dividends in the robustness, flexibility, and maintainability of the resulting code.

One of the real keys to effective software state-machine development is the proper documentation of the state machines themselves, which leads us to the more general topic of code documentation. *Internal code documentation*—i.e., documentation of the firmware within the source code itself—is often neglected or given extremely short shrift by programmers who feel they lack the time to include it. Inevitably, this lack of commenting is accompanied by the promise to do it "when the project's finished," a pledge that may or may not be fulfilled. Even if the comments are added after the firmware has been completed, they often lack the insight that can be imparted at the time the code is created, when the programmer is actually thinking about why the code is being written a certain way. In truth, failure to document the code at the time it is written is grossly irresponsible at best and is indicative of a programmer who is either too lazy or too self-serving (believing that a lack of comments ensures job security) to be entrusted with serious product development. Proper code documentation returns its time investment many times over by lessening the effort required to maintain the code in the future and by reducing the chance that future code changes will not account for assumptions implicit (but not documented). Any time someone tells you that commenting is unnecessary, you can be assured that the person truly does not understand how to develop software. It is imperative that programmers not succumb to the pressure to churn out undocumented code; it is, after all, their reputations that will suffer when (not if) problems arise because of poorly commented code.

With these general design principles as a foundation, we now turn our attention to the application framework itself. To aid our understanding, the examination will explore the framework from two different perspectives: the flow of data through the application and the temporal relationship between the various system elements. This gives the reader an appreciation both of the manner in which raw signal data is transformed into meaningful information and of the step-by-step operations required to implement that transformation. Our analysis begins with a study of the flow of data through the application.

Application Data Flow

Data flows through the application framework in a fairly straightforward manner, as shown in Figure 5.2. Raw analog sensor signals first pass through the sensor-specific analog signal-conditioning circuitry to bandlimit the frequency content of the signal to meet the Nyquist sampling criteria and to adjust the signal voltage to an appropriate range. The ADC module then digitizes the conditioned analog signals and stores the digitized data in a software buffer that is shared with the data-analysis software state machines. After processing the digitized data, the data analysis components transfer their results to a second shared buffer that serves as input for both the control-processing state machines and the data-reporting state machines. The control-processing modules update the control outputs according to application-specific algorithms, and data-reporting modules send the analyzed data to other external system components for use elsewhere.

Application Framework Data Flow

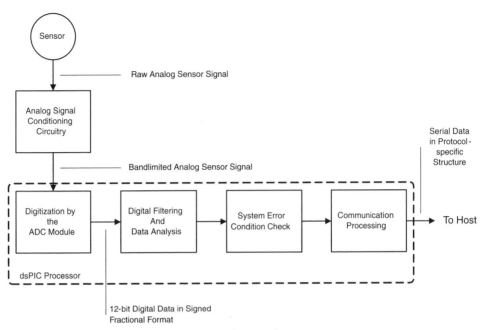

Figure 5.2. Generalized Application Framework Data Flow

All of this is performed in a lockstep fashion, with data entering the system at prescribed intervals and working its way through to the control algorithms and reporting interfaces before the next input sample. That's not to say that the control algorithms and reporting interfaces necessarily operate at the same sampling rate at which input data is sampled (frequently they don't), but there is a direct, usually constant, relationship between the input and the output rates. For example, an application might sample the input signals 1000 times per second (1 Ksps[6]) but produce analyzed data at a rate of only 250 samples per second, with the control algorithms and reporting interfaces operating at an even slower rate. The key point, though, is that each processing stage is updated at the output rate of the previous stage, and they must always be ready to handle the next sample as it becomes available.

There is one aspect of the data flow that is somewhat asynchronous, and that is the reporting of error conditions. These error conditions may reflect a problem with the input signal from the sensor (for instance, a signal value that is out of the expected range), an error with the basic hardware, an error in the algorithm, or merely a valid but undesired condition that is detected by the firmware (a cable is disconnected). Whatever the reason, the framework reports these anomalous operating conditions when they occur so that other elements of the system that depend upon the sensor's data know to take appropriate corrective action. The asynchronous nature of the error handling is indicated by the "error handler" lines emanating from each of the major processing modules and terminating at the reporting module.

Note that the communication link between the reporting module and the external system components may be any of the serial communication components: UART (for standard RS-232 or RS-485 links), CAN (Control Area Network), or SPI or I²C (for extremely short distances such as board-to-board). While the actual structure of the data reported will depend upon both the application's needs and the limitations of the communication channel, the high-level concept of reporting system data and anomalous conditions is the same for all communication modes.

System Task Flow

Having examined the flow of data through the framework, let's now look at the various software tasks needed to implement that data flow and the sequencing required for the tasks to operate properly. The implementation must carry out four essential tasks:

1. initialize the software environment, which in our case is the C run-time environment;

2. initialize the system hardware and software state machines to a known safe start-up state;

3. service the various interrupts as they occur and make their information available to the rest of the system in a coherent manner; and

4. operate an event-processing loop that continually checks for specific system events and performs the required processing when they occur.

Although specific elements of the framework may change from application to application (in particular the system initialization and the specific interrupt sources serviced), the general framework is a solid foundation on which to build the application. By using a framework, we can be more productive (since the framework is already tested), and we can devote more of our creative energy to the application (where we add value) and less to the underlying "plumbing."

One of the major issues with any application framework is the structure of the code, in particular ensuring that the code is easily customized for the needs of the specific application while still maintaining the modularity that insulates the rest of the system from those customizations. By using modular code, we can improve the stability of the system software since changes in one part of the software won't affect other areas of the code, and if (when) there are problems, the offending code can be tested relatively easily. The code in this book takes the approach of grouping routines that perform similar tasks (such as filtering or data analysis) and the data that they act upon in a single C source file along with a corresponding header file that contains the definitions and function prototypes required to use the routines. Doing so allows the designer to think of the system as a set of functional objects that communicate with each other through function calls and shared global data but that are otherwise "black boxes"[7] in the sense that the code that performs the data analysis, for instance, neither knows nor cares how the data filtering is performed; the analysis routines simply work with the filtered data.

The framework modules consist of the following ten major modules:

1. `Main.c / SystemEventDef.h`

   The application entry point main(), which calls the system initialization routines and then starts the event-processing loop.

2. `SystemCfg.c /SystemCfgDef.h`

   High-level system initialization routine that calls the low-level initialization routines for the ADC, filtering, data analysis, and communication modules, as well as routines to perform miscellaneous system tasks such as reading/ writing the system configuration data and resetting the watchdog timer.

3. `ADCIF.c / ADCIFDef.h`

   The analog-to-digital interface initialization, processing, and interrupt service routines that sample the sensor data.

4. `Filter.c / FilterDef.h`

   The filter initialization and high-level data-filtering routines.

5. `Analysis.c / AnalysisDef.h`

   The data-analysis routines that process the filtered sensor data to extract the desired information and that report it via the communication channels.

6. `CommIF.C / CommIFDef.h`

   The high-level wrapper routines that initialize the application-specific communication channels and control the flow of data through the channels.

7. `Timer.C / TimerDef.h`

   The timer initialization and interrupt service routines that perform time-critical processing tasks, that schedule data sampling, and that track elapsed system time.

8. `Protocol.c / Protocol.c`

   The high-level protocol handling routines that lie between the application and the communication routines in CommIF.c.

9. `Sensor.h / SensorDef.h`

   Routines for storing, retrieving, and applying sensor configuration information to the sampled data.

10. `dsPICDEMIF.c / dsPICDEMIF.h`

    Wrapper routines to easily access hardware resources on the dsPICDEM 1.1 General Purpose Development Board.

When implementing a given application, most of these modules will require some customization to work with the specific hardware platform and to perform the particular tasks required by that application. With this code structure, however, that customization can be handled in an incremental fashion that allows code changes to be tested thoroughly. For instance, the system initialization code is often the first code to be tested when bringing up a new hardware platform; the designer can implement the initialization code while leaving the other code stubbed out until the initialization code has been thoroughly tested. Once that code is working, the ADC interface can be brought to life and its operation verified, and so on until the entire application is up and going.

Should the hardware platform change between revisions, any required code changes would be limited to the `SystemCfg.c`, `ADCIF.c`, and `dsPICDEMIF.c` modules; if, on the other hand, we just want to extract different parameter values from the data, we only need to change `Analysis.c` to do so. Modularization as objects, with code and its associated data grouped together, can allow the designer to reuse significant portions of the software, reducing both development time and cost while increasing system reliability. All three benefits increase product profitability while making the designer's life easier, truly a win-win situation!

Initializing the Software Environment

Any program written in the C language must configure the C runtime environment immediately upon starting up, prior to executing any of the application code that begins with the `main()` routine. This initialization establishes important operating parameters such as the location of the stack and the heap[8], sets each variable in the various data sections generated by the compiler to its prescribed value, and then starts the application itself by calling the function `main()`. A list of the possible data sections is given in Table 5.1 along with the type of data stored in the sections.

Section Name	Description
.text	Not actually a "data" section in the usual sense, the .text section is the section that contains the executable code.
.data	One of the two initialized data sections that contain the information required to set the value of global variables that are explicitly initialized to a starting value. The .data section handles initialized data that has the far attribute and is the default section for all initialized data when the application uses the large memory model.
.ndata	The second of the two initialized data sections, .ndata is reserved for initialized data that has the near attribute and is the default section for initialized data when the application uses the small memory model.
.const	The .const section holds constant-valued data such as text strings or numeric data that has the const qualifier. Usually the section should reside in program memory and generally is accessed using the PSV window.
.dconst	Similar to the .const section in its contents, the .dconst section is created by the compiler under certain conditions when the application uses the large memory model.
.nconst	Also similar to the .const section in its contents, the .nconst section is created by the compiler under certain conditions when the application uses the small memory model.
.bss	The first of three sections that hold uninitialized data (global variables that are not initialized), .bss contains variables that have the far qualifier and is the default section for uninitialized data for applications that are built using the large memory model. Data in this section is cleared to 0 as part of the C runtime environment initialization.
.nbss	The second of three sections that hold uninitialized data, .nbss holds variables that have the near qualifier and is the default section for uninitialized data for applications that are built using the small memory model. As with the .bss section, this section is also cleared to 0 as part of the C runtime environment initialization.
.pbss	The final of the three sections that hold uninitialized data, .pbss is intended for RAM-based variables whose value should not be affected by a device reset (i.e., by the dsPIC DSC resetting). Unlike data in both the .bss and the .nbss sections, data in .pbss is not cleared to 0 or set to any other value during initialization. The .pbss section is located in near data memory.

Table 5.1. C30 Compiler-generated Code and Data Sections

Usually, the C runtime environment initialization code or *start-up* code is transparent to the application programmer, being incorporated in the software module crt0.o in the C30 compiler library file libpic30.a. If, for some reason, the designer would like to avoid initializing any of the initialized data sections (.data, .ndata, .const, .dconst, and .nconst), he can link in the file crt1.o instead of crt0.o; the only difference between the two is that crt1.o leaves out the data initialization step. Furthermore, if the designer requires any additional initialization that simply cannot wait until the application starts, he can edit the assembly-language file for either of the two start-up code modules (crt0.s and crt1.s, found in the src\pic30 subdirectory in the C30 compiler's directory) and include the requisite additions by simply linking in the new module.

It's important to note that the sample crt0.s and crt1.s source files supplied with the compiler are written specifically for the dsPIC30F2010, as indicated by the first two lines following the introductory comments:

...

```
        .equ __30F2010, 1
        .include "p30f2010.inc"
```

...

To ensure that the code is portable among the various devices in the dsPIC line, the Microchip C30 uses processor-specific include files that define the register address and bit assignments that are unique to each particular processor. As verification that the programmer is including the correct file for the desired processor, each include file has an assembler directive at the beginning of the definitions that checks whether the corresponding device ID has been defined and reports an error if it has not. In the sample code above, the first statement defines the device ID, and the second statement includes the processor-specific header file.

If the application targets a device other than the dsPIC30F2010, then the programmer has to indicate to the assembler the correct device for which the code is to be created. This task can be handled in any one of the following three ways, depending upon the designer's preference:

1. through a command-line switch when the assembler is invoked (if the programmer is using the command-line assembler), for example:

   ```
   C:\> AS -p30F6014A
   ```

2. by placing a .equ directive followed by the include file, as in:

   ```
       .equ __30F6014A, 1
       .include "p30f6014a.inc"
   ```

3. or by setting the processor in the MPLAB Configure → Select Device…dialog (the easiest way) as shown in Figure 5.3.

The final action of the C runtime initialization is to start the user's application code by calling the function main(), and it is to that routine that we now turn.

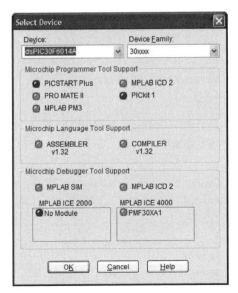

Figure 5.3. MPLAB Configure → Select Device Dialog

Initializing the System Hardware and Software State Machines

The start-up code intentionally performs the minimum initialization required to get the software environment up and running but does not do any application-specific configuration. Handling the initialization in this manner allows the C environment to be flexible and to be as platform-neutral as possible, but it also means that the programmer is responsible for configuring all of the dsPIC DSC's I/O ports and the on-chip peripherals. There is an initial, usually brief, time period during which the I/O signals are in their default reset condition, which may or may not be appropriate for the specific application. Depending upon the purpose of a particular signal, it might be necessary for the design to implement special hardware to ensure that the system operates safely even during the time between reset and programmatic initialization. If, for instance, one of the signals is used to turn on a high-power laser, we probably don't want it to pulse on start-up; even a short activation time could cause tremendous damage. It's important, therefore, that the first thing the application code does is to initialize the I/O ports to known-safe states.

Of course, for most systems I/O port configuration is just a small part of the overall system initialization procedure. The best approach to initialization is to begin with the I/O ports and then to work out in time-critical order by configuring the system (not just the dsPIC DSC) resources in the order that they will be needed. Depending on the specific environment, this may even mean that the application initializes certain off-chip components before setting up some on-chip modules;

the key is to get the system configured as quickly as possible and in the order that resources are required. The designer must also ensure that the system doesn't start processing system events until all of the resources required to process them have been successfully initialized. Ignoring this very important rule can result in erratic, even catastrophic, failure of the system.

Some designers prefer to initialize the hardware for all of the functional blocks prior to initializing the software state machines that work with those hardware components; others choose to initialize a software state machine immediately after configuring its associated hardware. The nature of the application itself may determine the choice of which approach to take; for instance, it may be critical to get all of the hardware modules to a certain state as quickly as possible. Where possible, it's helpful if one can take the module initialization approach in which the hardware and software for a particular functional element are initialized at the same time because this allows the code to be more modular, and it helps to ensure that complete system components are brought up in a well-defined order. In our case, the framework employs the modular initialization approach both for ease of coding and to clearly illustrate the ideas we're discussing.

dsPIC Interrupt Configuration

Configuring the interrupts is a four-step process:

1. identifying the interrupt sources to service,

2. setting the interrupt priority level (IPL) for each interrupt,

3. enabling the individual interrupts so they can be recognized by the interrupt controller, and

4. enabling the global interrupt so that the dsPIC DSC will process any of the interrupt sources that are both active and enabled.

While none of these steps are particularly difficult in and of themselves, they all three have to be performed correctly or the system won't operate reliably. The worst-case condition is if one or more of the steps is handled *almost* correctly, because that can lead to intermediate system failures that are extremely difficult to identify, replicate, and solve.

In our applications, we'll have three consistent interrupt sources: the system timers, the ADC, and the communication channels. Of these, the system timers and the ADC are the most critical because they sequence the flow of data through the system. The communication channel, although important, can survive operating at a lower priority because its functionality does not require the extremely tight time windows of the other two. If we temporarily have to buffer an incoming character, it's not too big a deal; if, on the other hand, we start to sample aperiodically, the system performance may degrade quickly.

The interrupt sources we'll need to service, in priority order, are:

Timers 2/3 – 32-bit timer mode, variable rate – ADC sampling timer

Timer 1 – 16-bit timer mode, 10-ms interrupt rate – general system timer

ADC Sample Ready – reports when a complete data sample is ready for filtering

UART 1 Rx Data Ready – receive data pending on communication channel

UART 1 Tx Holding Register Empty – transmitter available on communication channel

5.3 Implementation of the Framework Modules

Since we discussed the start-up code in some detail already, we'll begin our look at the framework implementation with the application entry point, the ubiquitous `main()` routine. As mentioned in Section 5.2, `main()` basically initializes the system components and then starts an event processing loop that continually checks for events that have been generated by either the interrupt service routines or by processing performed in the event loop itself. These system events are really just bit-mapped flags in the global 16-bit variable `g_vui16SysEvent`,[9] with the events defined in the header file `SystemEventDef.h`. Setting a flag indicates that the event has occurred and needs to be processed; clearing it shows that the event has been handled.

The following code for `main()` shows the call to the system initialization routine: `SystemInit()`:

```
Int16
  main(void)
  {
// Local Variables

  Uint8
    ui8Analysiscount,     // Decimation count for scheduling data analysis
    ui8RxData,            // Received communication data
    ui8Status;            // Function execution status

// Initialize the system hardware and the
// associated software state machines

  SystemInit();
```

```
    //  Process system events as they occur

while (FOREVER)
  {
  //  Are there any pending system events? Start
  //  by checking for sampled data that is ready
  //  to be filtered

  if (g_vui16SysEvent)
    {
    //  Yes, is sampled data ready to be filtered?

    if (g_vui16SysEvent & EVT_FILTER)
      {
      //  Yes, data samples are ready so clear
      //  the event and filter the samples

      g_vui16SysEvent &= ~EVT_FILTER;
      ui8Status        = FilterData();

      g_vui16SysEvent |= EVT_ANALYZE; // Yes, signal
                                      //   Analyze Data
                                      //   event

      }

    //  Is there filtered data to be analyzed?

    if (g_vui16SysEvent & EVT_ANALYZE)
      {
      //  Yes, so clear the event and
      //  perform the analysis

      g_vui16SysEvent &= ~EVT_ANALYZE;
      ui8Status        = AnalyzeData();

// Did the analysis reveal a condition
// that should be reported?

ui8AnalysisCount++;      // Increment the data analysis decimation count

if (ui8AnalysisCount >= ANALYSIS_TIME)
    {
    ui8AnalysisCount = 0;               // Reset the decimation count
    g_vui16SysEvent |= EVT_REPT_RESULTS; // Flag that we have results
                                        //   to report

      }
```

```
      //  Are there analysis results to
      //  be reported to the host?

      if (g_vui16SysEvent & EVT_REPT_RESULTS)
        {
        //  Yes, so clear the event and report
        //  the results via the communication
        //  channel

        FormatResultsMsg(g_ui8ResultsMsg,
                      &g_ui16ResultsMsgLength);

        if (g_ui16ResultsMsgLength <= CommGetTxFreeCount())
          {
          //  Have room in the transmit queue so add
          //  in the results message and clear the
          //  event to show that we've processed it

          CommPutBuff(g_ui8ResultsMsg,
                    g_ui16ResultsMsgLength);
          g_vui16SysEvent &= ~EVT_REPT_RESULTS;
          }
        }

    //  Has a 100 msec timer tick occurred?

    if (g_vui16SysEvent & EVT_TIMER)
      {
      //  Yes, clear the event and perform any
      //  required timer tick processing

      g_vui16SysEvent &= ~EVT_TIMER;

      //  Insert any additional processing
      //  to be performed here

      }

//  Check whether we have received
//  any data from the host

if (CommGetRxPendingCount() > 0)
  {
  //  Yes we do, so get the next received
  //  character and process it
```

```
    ui8Status = CommGetRxChar(&ui8RxData);

    if (ui8Status == ST_OK)
        CommProcRxChar(ui8RxData, &g_ui8CommParseState);
    }

// Reset the watchdog timer
// to keep it from expiring

ResetWatchdogTimer();
    }

// We should NEVER get here if the system
// is operating normally. The routine will
// exit and the startup code will reset the
// device if we get to this point.

return ST_SYSTEM_FAIL;
}
```

Two things should be noted about the system initialization: it occurs before anything other than allocating the local variables, and the details of the initialization are wrapped in a separate function rather than included as part of `main()` itself. The first ensures that the hardware is initialized to a known safe state as quickly as possible, while the second is an example of writing good, modular code that is relatively platform neutral (at least at the high level).

We'll look at the system initialization code in greater depth shortly but for now let's continue with `main()`. After the system's been initialized, the code starts a processing loop that continually checks for and processes system events, checks for and processes any data received over the communication channel, and resets the watchdog. The looping continues until the processor is reset.

As implemented, the loop can check and process multiple events during a single pass through the loop. Depending upon the worst-case time it would take to process every possible event and to check and process any received communication data, the loop structure may have to be changed to process only a single event for each pass through the loop. Such a change is easily made by simply changing the series of `if()` statements to `if()`...`else if()` statements. The one criteria that must be met is that the watchdog timer must be reset before it expires, or the entire device will reset. Usually, we only want that to occur if the software becomes stuck in an invalid processing state, not just because we took a bit too long to go through the main processing loop.

To get a better feel for how we'll perform the system initialization, let's examine the `SystemInit()` routine more closely. The configuration code uses functions found in the dsPIC30F DSC Peripheral Library,[10] which is included with the C30 compiler and is also available on the Microchip website. Documentation for the library is found in the *16-Bit Language Tools Libraries* document[11] located in the `MPLAB C30\Docs` subdirectory of the directory in which the compiler is installed, or in the same document on Microchip's website. Those functions are further wrapped (albeit lightly) in code that is application specific so that, should the underlying hardware or software platform change (for instance, should a better library become available to implement the desired functionality), the firmware components could be swapped out fairly easily with new code.

One other key aspect of all of our systems is their ability to perform advanced digital filtering using the Microchip DSP Library routines, and in most cases we'll want to use FIR filters. To perform this filtering, the application must first create and initialize a data structure for each filter. This `FIRStruct` structure, whose programmatic definition is found in the compiler-supplied header file `dsp.h`, contains all of the information required to maintain a filter's tapped delay line and to apply the filter coefficients. Creation of the structure is simply a matter of allocating both the filter coefficients vector (and initializing it with the coefficient values) and the vector for the tapped delay line), then either declaring the FIR structure as a global variable or alternatively declaring it as a local variable within a function whose scope is active for as long as the filter needs to be in existence. As a rule, the author prefers the former approach because even if the execution paths for the application code change, the filter will always be in scope. If, on the other hand, the filter is declared as a local variable and the execution paths are changed in a future software version, the filter structure may temporarily go out of scope, allowing other data to overwrite the memory structure and causing erroneous results when the filter structure data is used again.

One of the great aspects of the dsPIC Filter Design software is that it will generate assembly-language code that handles creation of the filter structure and its associated tapped delay line and filter coefficients at the touch of a button. For example, to create the filtering code for our thermocouple application, we perform the following steps in Filter Design:

1. From the main menu toolbar, select **Design→FIR Window Design...** as shown in Figure 5.4.

Figure 5.4. FIR Filter Design Menu Selection

2. The program will then display the first filter design window, which is shown in Figure 5.5. Select the **Lowpass** filter option and press the **Next** button.

Figure 5.5. FIR Filter Design Window 1 of 4

3. Enter the filter parameters as shown in Figure 5.6, and then click on the **Next** button to move to the third design window.

4. Select the type of filter that you would like to use, noting the number of taps required for each filter type in order to implement the filtering requirements specified in the previous window. In Figure 5.7, we select the Gaussian filter for our application, and it will require 51 taps. Note that we could have also specified a particular number of taps if we so desired.

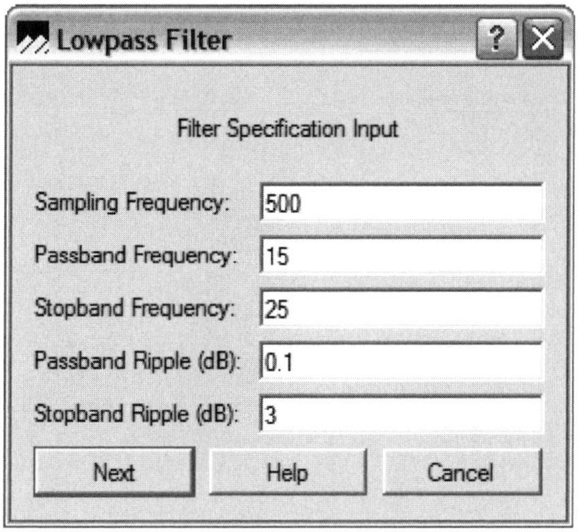

Figure 5.6. FIR Filter Design Window 2 of 4

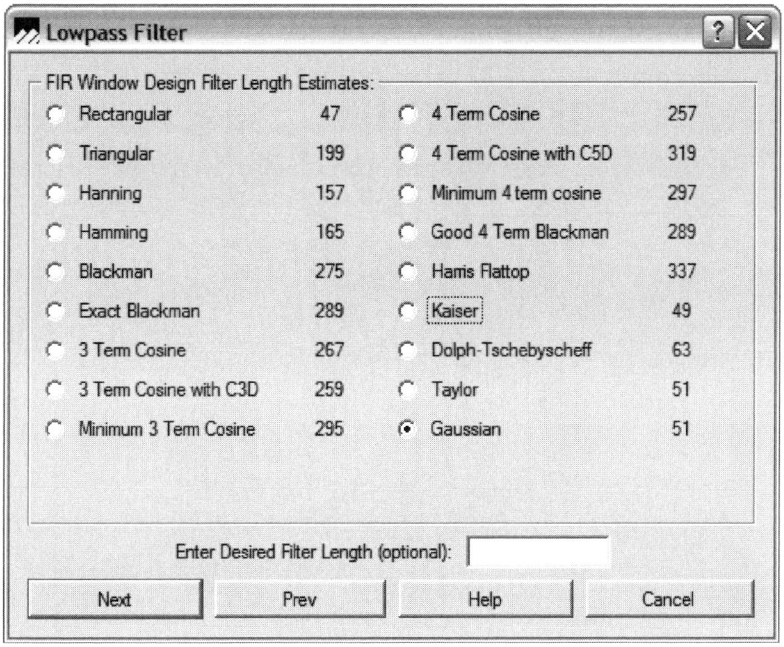

Figure 5.7. FIR Filter Design Window 3 of 4

5. After pressing the **Next** button in the screen in Figure 5.7, the program will generate the resulting filter response curves that are shown in Figure 5.8. At this point, we've created the filter, but we haven't generated any code.

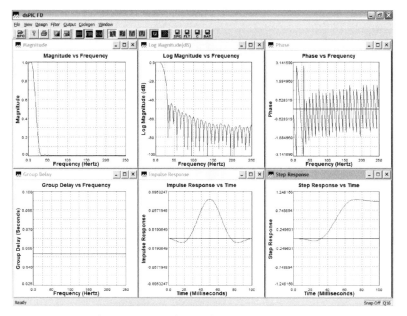

Figure 5.8. FIR Filter Design Window 4 of 4

6. To generate the filter code itself, select the **CodeGen** → **Microchip** → **dsPIC30** entry from the main menu toolbar as shown in Figure 5.9.

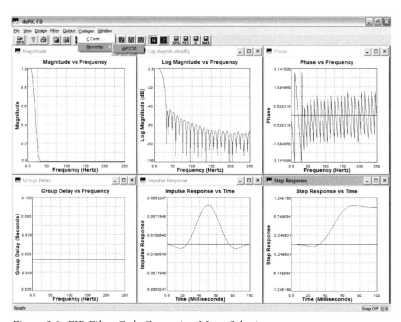

Figure 5.9. FIR Filter Code Generation Menu Selection

7. The program will display a dialog box (Figure 5.10) that allows the user to enter the desired code-generation options. In our case, we'll use the default selections **(Use General Subroutine and X Data Space)** and add one more, the **C Header File and Sample Calling Sequence (.h)** option that will tell the software to generate an associated C header file.

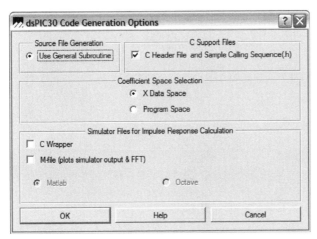

Figure 5.10. Code Generation Options Dialog

8. Pressing the **OK** button in the dialog box will display the **dsPIC30 code base file name** dialog shown in Figure 5.11 that lets the user specify the name of the generated code file. Because the code generator uses the file name as the base name for the created FIR filter structure, the file name must not have any spaces or punctuation.

   Enter the file name SensorFilter and press the **OK** button.

9. The program generates a number of files, the most important for us being the files `SensorFilter.h` (the C header file that defines our filter structure for use by other C files) and `SensorFilter.s`, the assembly-language code containing the filter structure and its associated coefficient and tapped delay-line buffers.

To actually use the generated filter code in the application, the programmer simply adds the filter code (in this case, the file `SensorFilter.s`) to the project. Before the filter is first employed by the application, it must be initialized by calling the function `FIRDelayInit()`, which initializes the filter's tapped delay line to a known state. Note that if code generated by the Filter Design package is being used, there is no need to call the function `FIRStructInit()` to initialize the associated

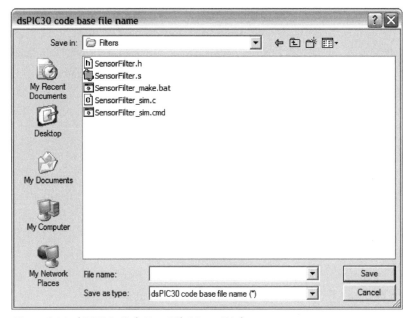

Figure 5.11. dsPIC30 Code Base File Name Dialog

structure; that task is handled by the generated code. Failure to initialize the tapped delay line will cause problems with the filter's response until new data is able to work its way through the delay line.

Once the filter has been initialized, applying it to new data is extremely easy. When a block of new data is available to filter, the application calls the function FIR() with pointers to both the new data to filter and to the destination data buffer that will hold the resulting filtered values, as well as a pointer to the filter structure being applied. Since FIR() handles updating of the associated tapped delay line, the application does not need to concern itself with that task.

There is one final caution about using the DSP library routines, and it is an important one. If a library routine is interrupted, the interrupting routine must ensure that it restores the contents of the Status register and the DO and REPEAT instruction values to the pre-interrupt state, or the DSP function may return invalid results. The DSP routines use certain shared hardware resources, and if an interrupt routine changes the configuration of those resources, the change will probably cause problems with the portion of the DSP function that executes after the interrupt completes. To make things easier for the designer, the description for each of the DSP library routines lists the resources it uses.

5.4   Summary

This chapter has provided an overview of the software framework we'll employ in the next three chapters, all of which are real-world applications of the principles that we've been discussing to this point. Although much of the framework is somewhat skeletal at this point, we'll be adding flesh to those bones quickly, starting with the temperature sensor that we'll develop in the next chapter.

Endnotes

1. The Microchip part number for the dsPICDEM 1.1 General Purpose Development Board is DM300014.

2. The Microchip website is www.microchip.com. The student edition is full-featured for the first 60 days following installation, after which it supports only minimal code optimization. In most cases, optimization at this level will cause the code size to grow, but the compiler functionality is otherwise unchanged. The part number for the full-featured version of the compiler is SW006012.

3. The Microchip part number for the ICD2 is DV164005.

4. The Microchip part number for the MPLAB IDE is SW007002.

5. As with anything associated with development tools, the wise designer takes the term "tested code" with a grain of salt. Although programmers as a class have a tendency to immediately assume that "it's the other guy's code that has the problem," a better approach is to presume that one's own untested code probably is the culprit when an error occurs. If that code has been examined thoroughly and found to be error-free, however, one then should verify that the third-party "tested" software is indeed operating properly (after taking the all-too-often ignored step of confirming that the parameters to the tested code are correct). Remember, just because software is "tested" doesn't mean that it was tested for the specific condition that causes a problem.

6. 1 Ksps = 1 kilosamples per second = 1,000 samples per second

7. The term black box refers to a system in which we know how we expect the outputs to behave given a particular set of input values, but do not know (nor particularly care about) the internal operation of the system. An example of a black box system for many people would be the ignition system of their car.

They know that if they turn the ignition key, the car should start; how that actually occurs is frequently a mystery and not something that they think about.

8. The *stack* and the *heap* are two areas of memory that are established at runtime (i.e., when the program first starts. The stack is a block of memory that is used to hold temporary variables (that is, the local variables for a function), to store parameters passed to a function, and to hold the program address to which to return when a function has completed. During the time the application runs, the stack may grow and contract as necessary.

9. The variable `g_vui16SysEvent` illustrates the variable-naming convention used throughout the code. In general, variable names are prepended with their abbreviated type; for instance, `vui16` would indicate a volatile unsigned 16-bit value. Global variables are further prepended by a `g_` to indicate that they are global, rather than local, variables.

10. The dsPIC30F DSC Peripheral Library is Microchip part number SW300021.

11. The *16-Bit Language Tools Libraries* manual is Microchip document DS51456C.

6

Sensor Application—Temperature Sensor

It doesn't make a difference what temperature a room is.
It's always room temperature.

—Steven Wright

Each of the next three chapters (Chapters 6 through 8) develops a complete sensor system that measures some common parameter and then communicates the measurements to a host system. This chapter tackles temperature measurement, Chapter 7 constructs a pressure and load monitor, and Chapter 8 creates a flow sensor. By necessity, each application focuses on a particular type of sensor, but the concepts apply to a broad range of sensing elements, often with only minor adjustments needed to address the unique requirements of the specific sensor being employed to measure the particular parameter of interest. The reader should feel free to expand upon the ideas presented here; one obvious extension would be to include a control algorithm that acts upon the sensor's parameter measurements to accomplish a desired effect. With that introduction, let's turn our attention to temperature, probably the most widely measured physical property in the world today.[1]

Temperature's importance comes from its impact on so many environmental and processing situations. We, or more accurately the equipment we employ, require accurate temperature measurements to heat and cool our homes, to operate our car's engine, to cook food, to process industrial materials, to monitor a patient's vital signs, and the list goes on and on. Back in Chapter 1, we saw an example of such a device, the mercury bulb thermometer, in which the absolute temperature is indicated by the level of the liquid mercury inside the thermometer. Unfortunately, the fragility of that type of sensor (usually a glass tube enclosing the column of mercury) and the toxicity of its sensing element (mercury is highly poisonous) limit the mercury thermometer's usage to certain well-controlled environments. The sensor system we develop here employs a sensing element that is far better suited to a wide variety of environments and applications: the thermocouple. Before we delve too deeply into the details of thermocouples, let's first look at the variety of temperature-sensing elements available to us.

6.1 Types of Temperature Sensors

Over the years, scientists have developed a host of specialized sensing elements that respond to absolute temperature or to changes in temperature by varying some physical property. In this book, we're interested primarily in sensing elements that we can monitor using electronics, for the very simple reason that doing so allows us to easily convert the monitored physical property into a measurement that we can manipulate using the dsPIC DSC. That's not to say that other approaches (such as visually monitoring the mercury bulb thermometer) are invalid; it merely means that those nonelectrical techniques are unsuitable for our purposes because of the monitoring platform. Nails and screws can both be used to fasten two boards together, but one is far more likely to use a nail if the tool at hand is a hammer.

Currently, there are five widely used types of sensing elements whose outputs can be monitored electrically. Advances in technology are sure to build upon this list, but most electrically based temperature sensors are one of the following types:

1. thermocouples,

2. resistance temperature detectors (RTDs),

3. thermistors,

4. silicon sensors, or

5. infrared sensors.

Although we'll discuss each of these sensor types in the following sections, the application we develop in this chapter uses the thermocouple exclusively because thermocouples are widely used, well understood, accurate (when utilized properly) and relatively inexpensive (which is one reason they're so widely used).

Thermocouples

Thermocouples are two-wire sensing elements that make use of the *Seebeck effect* to measure the temperature of the junction of the two wires. The Seebeck effect, discovered by the scientist Thomas Seebeck in 1821,[2] creates a voltage across the junction of any two dissimilar metals that correlates to the temperature of the junction. Although this voltage is quite small, on the order of several microvolts per degree of temperature, it's possible to create systems that are accurate over a wide temperature range provided that proper analog and digital signal processing techniques are used.

The caveat "that proper analog and digital signal processing techniques are used" is a major consideration. The very small voltages produced by thermocouples (on the order of millivolts) require that designs employ good grounding and shielding techniques to avoid introducing unacceptable levels of noise in the measured voltage. In addition, because the traces on a printed circuit board are made of a metal that differs from those of the thermocouple, the very circuitry that we use to measure the original thermocouple voltage introduces an additional Seebeck junction whose output varies with temperature! Finally, thermocouples can be highly nonlinear over their range of measurement, as we can see from Figure 6.1, which shows the response of various thermocouples with temperature.

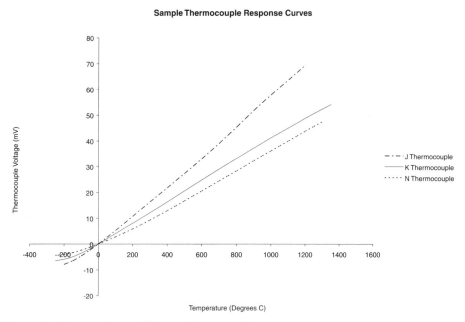

Figure 6.1. Response Curves of Various Thermocouples

Given all these serious constraints, why would anybody select a thermocouple to measure temperature? Thermocouples are popular for three reasons: they are relatively inexpensive, they work over a wide range of temperatures, and we can compensate for their limitations (as we'll see in the Challenges topic in Section 6.2). Even with the additional circuitry and software needed to deal with thermocouples' deficiencies, they make a reliable sensor whose cost and performance characteristics satisfy a large number of applications.

Resistance Temperature Detectors (RTDs)

Another very popular temperature sensor is the *resistance temperature detector* or *RTD*. Usually constructed of fine (small diameter) platinum wire wound around a nonelectrically conductive cylinder or *mandrel* with the entire assembly being coated by a nonconductive material, the resistance of the RTD varies linearly with temperature. By passing a known current through the device, we can generate an output voltage that corresponds to the temperature of the RTD through the equation:

$$R_{\mathrm{RTD}} = R_0 + \alpha T_{\mathrm{RTD}}$$

where

R_{RTD} equals the measured resistance in ohms of the RTD at temperature T_{RTD}

R_0 equals the resistance in ohms of the RTD at 0°C

α equals the temperature coefficient of the RTD

T_{RTD} equals the temperature of the RTD in °C

The temperature coefficient of an RTD depends upon the purity and composition of the platinum used to wind it. Two standard RTDs are widely used, one having an α value of 0.00392 and the other a slightly lower α value of 0.00385. Because RTDs with the lower α value were initially used extensively in Europe, devices with that value are said to have a *European curve*, while those with the higher value are said to possess an *American curve.*

Since RTDs are far more linear, why would a designer ever choose to employ a thermocouple instead of an RTD? There are two aspects to the answer, one economic and the other having to do with system performance. RTDs tend to be more expensive than their thermocouple counterparts, so the initial device cost may be more expensive. Often, the more serious constraints arise in the area of system performance.

We usually want to measure the temperature of the system without disturbing the system itself. Because an RTD requires current to operate (necessary to generate a voltage across the device), there will be some nonzero amount of sensor element self-heating due to the current flow. This effect is particularly noticeable in applications in which there is little flow of material past the sensor because the added heat builds up, causing a thermal error that may change over time. The thermal error is real, in the sense that the surrounding material actually becomes warmer, but it may not accurately reflect the temperature of the material or environment even a short distance from the sensor. Thermocouples, which are essentially zero-current devices, suffer from no such self-heating effect.

The other system-performance constraint has to do with the temperature range over which RTDs operate, which is significantly narrower than the range for thermocouples as a class (although individual thermocouple types may have a range limitation similar to RTDs). RTDs work for temperatures between –250°C and +850°C, whereas thermocouples have an operating range of –270°C to +2300°C. Particularly for high-temperature applications above 850°C, thermocouples may be the only viable choice of sensing element.

Thermistors

Like RTDs, thermistors change their device resistance with temperature, but unlike RTDs, whose resistance rises with temperature, the resistance of a thermistor actually *decreases* with increased temperature. This *negative temperature coefficient* effect is only one way in which the two devices differ because, unlike RTDs, thermistors are also highly nonlinear. Fortunately, thermistors can be used in matched pairs so that the nonlinearities of one thermistor compensates for the nonlinearities of the second thermistor, resulting in a reasonably linear output.

Because they require a current to operate, thermistors share RTDs' self-heating problem, and their operating range is significantly less than that for either RTDs or thermocouples, with a typical operating range being from –40°C to +150°C. Within this range, though, properly configured thermistors can be extremely accurate.[3]

Silicon Sensors

It's not at all unusual to include a temperature sensor in electronic systems that have to operate in harsh environments. The system can use the measured temperature to either activate climate control or to shut down if things get too hot or too cold for it to run safely, or it can use the temperature reading to perform other application-specific tasks. Whatever the reason, the increasing integration of system components through the years led inexorably to the creation of first silicon-based temperature sensors and then the integration of those sensors into other chips.

In general, silicon temperature sensors may be fairly accurate (on the order of RTDs or thermistors), but they usually are limited to an even more restricted range of operation than thermistors (–40°C to +125°C), and they suffer from the same self-heating problems as any sensing element that requires power to operate. Silicon sensor devices themselves tend to be more expensive than their nonsilicon counterparts, but they often don't have additional wiring costs associated with them since they frequently are collocated with the electronics they are monitoring. Even if they are located some distance from the processing unit, the wiring required to connect them to the main system can be standard copper wire rather than expensive thermocouple or RTD wiring.

Infrared Sensors

Some temperature-measurement applications preclude the use of a sensor that is in physical contact with the material being monitored. Examples of this would include extremely hot substances (above the 2300°C limit for thermocouples) or materials that would be adversely affected by touch, such as a thin film or a painted surface prior to drying. In cases like this, one approach is to measure the emitted infrared radiation of the object and then to compute the corresponding temperature based on the object's emitted electromagnetic energy.

While simple in concept, infrared temperature sensing can be difficult in practice for a number of reasons. The primary difficulty is that different materials have different *emissivity*[4] characteristics—i.e., they emit infrared radiation with varying efficiency. If we are to use an infrared sensor properly, we have to know the emissivity characteristics of the material being measured or the temperature readings will be inaccurate. An example of this effect in the real world can be seen when monitoring the temperature of molten plastic in the metallic barrel that channels the plastic into a mold. Unless the operator knows what he or she is doing, it's not at all unusual for the measured temperature to be that of the metallic barrel rather than the desired temperature of the plastic melt.[5]

A second issue that may be either an advantage or a problem, depending upon the circumstances, is that infrared temperature measurements are made over an area rather than at a single point. While this is true to a certain extent with all temperature sensors (after all, we have yet to invent a sensor that takes up zero area), the other sensors that we've discussed tend to be treated as point-source measurements—i.e., they are usually small enough that we can ignore the size of the sensor itself. This is not the case with infrared sensors, which inherently examine a projected area on the object being measured, essentially reporting an average temperature over that area. When using such a sensor, it's important that the operator completely fill the sensing area with the object, or the sensor will report inaccurate readings that average in surrounding temperatures.

A third potential problem with infrared sensors is that they can pick up reflected infrared radiation from other sources as well as that emitted from the target. Like all electromagnetic waves, infrared radiation will reflect off of any surface with which it comes in contact, so some of the measured infrared energy may well come from other sources. If the reflected energy is quite small relative to the emitted energy that's picked up by the sensor, then the effect may be negligible, but if the reflected energy is significant, it can seriously degrade the quality of the readings.

One of the biggest issues with infrared sensing, though, is its cost. It's not unusual for an infrared sensor to cost multiple thousands of dollars, several orders of magnitude more than any of the other sensors discussed. With a price tag like that, with the additional training required to use it, and with the added operational complexities (in particular, needing to have detailed information on the infrared radiation characteristics of the substance being monitored), infrared sensing is currently limited in its application.

6.2   Key Aspects of Temperature Measurement

In any temperature-measurement system, there are certain fundamental issues that we have to address in order to get meaningful, accurate results. Failure to consider any of these issues can lead to inaccurate measurements. Depending on the particular situation, these inaccuracies may range from the inconsequential (if your air conditioning thermostat is off, you'll simply adjust it to get the desired conditions), to the frustrating (try baking a cake at what you think is 400°F when the oven is actually 250°F[6]), to the catastrophic (imagine a hydrogen gas production line that becomes overheated).

For the rest of this chapter, when we discuss a particular issue, we'll concentrate on how that issue affects thermocouple sensors, but the subjects we'll explore must be addressed in any temperature-measurement system regardless of the type of sensor used. To ensure that our system works properly, we'll consider the following important topics:

1. the types of temperature sensors available,
2. the required measurement range,
3. the resolution and accuracy we need (the two are not equivalent),
4. the characteristics of the thermocouple signal, and
5. the sources of noise in the measured signal.

Range of Measurement

In any measurement system, we first must identify the range of values that we need to be able to process, because that frequently determines the type of sensor that we can use. Although thermocouples as a class can be used to measure temperatures from –270°C to +1760°C, individual thermocouple types cover only a portion of that range, as shown in Table 6.1, which is taken from the National Institute of Standards and Technology table of thermoelectric voltages and coefficients.[7]

Type	Range (°C)
B	0 – 1820
E	−270 – 1000
J	−210 – 1200
K	−270 – 1370
N	−270 – 1300
R	−50 – 1760
S	−50 – 1760
T	−270 – 400

Table 6.1. Thermocouple Measurement Ranges

The measurement ranges given in the table are really just a starting point, however, because they represent the maximum conditions under which a particular thermocouple type can be used. A thermocouple's actual operating range is then derated based on its wire diameter (the smaller the wire, the more restricted its temperature range) and the temperature rating of the protective sheathing material around it (if any). As an example of how severe these derating effects can be, the upper limit for a J thermocouple in the table is given as 1200°C, but checking the reference table for a commercially available protected bare-wire J thermocouple[8] shows that an 8 AWG (0.128" diameter) thermocouple is rated to only 760°C. If the thermocouple is constructed of 36 AWG (0.005" diameter) wire, the maximum thermocouple temperature drops to only 315°C, a loss of nearly 75% of the expected temperature range.

One other real-world factor may affect the selection of the thermocouple type, namely that the designer may not have a "choice" at all. It's not at all uncommon for an end-user to already have a standard thermocouple specified, and changing that standard may not be an option. In North America, for instance, most injection-molding operations use J-type thermocouples while European and Asian customers tend to use K-type thermocouples for the exact same applications. The wise system designer makes provisions for supporting more than one type of thermocouple if at all possible; it saves both the end-user and the product manufacturer time and trouble.

Resolution of Measurement

The required measurement range not only determines the type of thermocouple we need to use, it also affects the *resolution* we can expect to get in our measurements. Resolution refers to how finely we perform our measurements and is usually expressed in terms of degrees. A fairly common requirement is to be able to resolve our temperature measurements to a half a degree C or perhaps a degree F, but we are able to do so only if our ADC can digitize the corresponding analog sensor signal with sufficient resolution. In order for that to happen, two conditions must hold:

1. we must be able to transform the analog output signal range from the thermocouple into an analog voltage range that can be digitized by the ADC, and

2. the ADC must be able to digitize the transformed signal range with the desired resolution.

While we've extolled the benefits of processing the sensor signal in software, the first condition is one that we can satisfy only through electronic hardware, usually by placing amplification and level-shifting circuitry between the sensor output and the input to the dsPIC's ADC. If we fail to properly map the sensor output signal levels to a suitable voltage range for the ADC, the digitized signal might clip[9] and might even destroy the dsPIC DSC if the input exceeds the absolute voltage input range specified by the chip's data sheet. Adding an amplification stage also allows us to buffer the sensor's output signal, which typically has a fairly limited drive capability, from the dsPIC DSC's relatively low-impedance input to the ADC. By inserting the buffer, we keep the dsPIC's ADC from introducing unwanted signal distortion due to excessive loading of the sensor output signal.[10]

How, then, does the designer accomplish this transformation? The procedure is simple and involves just three very basic calculations. Before making these computations, though, we need to know the voltage range for the sensor's output signal (at least for the measured parameter values of interest) and we need to know the ADC's input voltage range. Armed with this knowledge, we then perform the following three calculations:

1. Express both the sensor's output signal and the ADC's input voltage ranges in terms of their spans and offsets:

$$\text{SPAN}_{\text{SENSOR}} = \text{VS}_{\text{MAX}} - \text{VS}_{\text{MIN}}$$

$$\text{OFFSET}_{\text{SENSOR}} = \text{VS}_{\text{MIN}}$$

$$\text{SPAN}_{\text{ADC}} = \text{VA}_{\text{MAX}} - \text{VA}_{\text{MIN}}$$

$$\text{OFFSET}_{\text{ADC}} = \text{VA}_{\text{MIN}}$$

where

$\text{SPAN}_{\text{SENSOR}}$	is the sensor output voltage span
VS_{MAX}	is the maximum sensor output voltage
VS_{MIN}	is the minimum sensor output voltage
$\text{OFFSET}_{\text{SENSOR}}$	is the sensor output voltage offset
SPAN_{ADC}	is the ADC input voltage span
VA_{MAX}	is the maximum ADC input voltage
VA_{MAX}	is the minimum ADC input voltage
$\text{OFFSET}_{\text{ADC}}$	is the ADC input voltage span

2. Compute the amplification gain required to map the sensor output voltage span to the ADC's input voltage span:

$$\text{GAIN} = \text{SPAN}_{\text{ADC}} \,/\, \text{SPAN}_{\text{SENSOR}}$$

where GAIN is the required amplification gain. If the gain is greater than 1, the sensor output signal is amplified in the traditional sense (i.e., it's made larger); if the gain is less than 1, the sensor's output is attenuated.

3. Compute the level shifting required to map the sensor's voltage offset to the ADC's required minimum voltage:

$$\text{SHIFT} = \text{OFFSET}_{\text{ADC}} - \text{OFFSET}_{\text{SENSOR}}$$

where SHIFT is the voltage that must be added to the sensor output voltage *after* amplification.

Having satisfied the first of our two conditions (mapping the output sensor voltage range to an appropriate ADC input voltage range), we must now examine the steps required to fulfill the second condition (verifying that the ADC can digitize the transformed voltage range with sufficient precision). This, too, is quite straightforward and is another three-step process:

1. Compute the range of the measured parameter values that correspond to the sensor output voltage range from the previous set of calculations:

$$\text{PARM}_{\text{RANGE}} = \text{PARM}_{\text{MAX}} - \text{PARM}_{\text{MIN}}$$

where

$\text{PARM}_{\text{RANGE}}$	is the range of parameter values that correspond to the sensor output voltage range, expressed in units of measurement

appropriate for the parameter and not in terms of the corresponding voltage itself

$PARM_{MAX}$ is the parameter value that corresponds to the maximum sensor output voltage (note that this is *not* the maximum sensor output voltage but rather the parameter value that corresponds to that voltage)

$PARM_{MIN}$ is the parameter value that corresponds to the minimum sensor output voltage (as with $PARM_{MAX}$, this is the parameter value, not the sensor output voltage value)

2. Compute the resolution of the ADC in counts—i.e., determine the total number of levels into which the ADC input voltage range can be divided given the bit-resolution of the ADC:

$$RES_{ADC} = 2^{NumBits}$$

where

RES_{ADC} is the resolution of the ADC expressed as the number of counts, or levels, into which the ADC can divide its input voltage range

NumBits   is the number of bits that the ADC uses for digitization (either 10 or 12, depending upon the type of dsPIC DSC)

3. Calculate the resolution in terms of the measured parameter, expressed as the parameter units per ADC count or level:

$$PARM_{RES} = PARM_{RANGE} / RES_{ADC}$$

where

$PARM_{RES}$ is the parameter resolution, expressed as the parameter units per ADC count or level

A simple example should help clarify these concepts. Suppose that we want to use a J-type thermocouple to measure temperatures in the range 0°C to 750°C, and further assume that we select a thermocouple constructed to handle the required temperature range so we can concentrate on just the signal-processing needs here. From the ITS-90 table of thermoelectric voltages for a J-type thermocouple, we can see that a J thermocouple produces a voltage of 0.0 mV at 0°C and a voltage of 42.281 mV at 750°C. For ease of scaling, let's assume an input voltage range of 0 mV to 50 mV (approximately 870°C) and assume that we're running our dsPIC ADC with an input range of 5V. The gain and level shift required for our amplification stage are found by:

1. Computing the span and offset of the sensor output voltage and the ADC input voltage:

$$\text{SPAN}_{\text{SENSOR}} = 0.050V - 0.000V = 0.050V$$

$$\text{OFFSET}_{\text{SENSOR}} = 0.000V$$

$$\text{SPAN}_{\text{ADC}} = 5V - 0.0V = 5V$$

$$\text{OFFSET}_{\text{ADC}} = 0.0V$$

2. Calculating the amplification gain required to map the output sensor voltage span to the ADC's input voltage span:

$$\text{GAIN} = 5V / 0.050V = 100$$

3. Calculating the level shift required to map the minimum output sensor voltage to the corresponding minimum ADC input voltage:

$$\text{SHIFT} = 0.0V - 0.000V = 0.0V$$

In this case, since the analog input voltage range and the analog output voltage range both have a lower limit of 0V, no voltage offset is requred in the analog amplifier stage. In the general case, however, the amplifier stage requires a level-shifting component as well to map the lower limit of the input voltage range to the desired lower limit of the output voltage range.

With the input voltage range mapped, we're ready to verify that the second condition is valid. If our dsPIC DSC uses a 12-bit ADC, each bit of the ADC output represents 1/4096 of the ADC input voltage range ($1 / 2^{12} = 1 / 4096$). In our example, this means that we're able to digitize the amplified temperature signal with a resolution of:

$$T_{\text{RES}} = T_{\text{RANGE}} / \text{RES}_{\text{ADC}}$$

$$T_{\text{RES}} = (870°C - 0°C) / (4096)$$

$$T_{\text{RES}} = 0.2124°C$$

where

T_{RES}	is the temperature resolution of the system in °C
T_{RANGE}	is the temperature input range of the system in °C
RES_{ADC}	is the resolution of the ADC in counts

A quick calculation shows that if we used a dsPIC DSC with a 10-bit ADC (which has a resolution of $2^{10} = 1024$ counts), our temperature resolution for this example would degrade to:

$$T_{RES} = (870°C – 0°C) / (1024)$$
$$T_{RES} = 0.8496°C$$

If we want to maintain a temperature resolution of 1°F (0.56°C), we would be able to do so with the 12-bit ADC, but the 10-bit ADC would fail to meet our system requirements.

Accuracy of Measurement

Frequently users, and sometimes designers as well, confuse a system's resolution with its *accuracy*, which is a totally different beast. While resolution tells us the degree of granularity to which we can compute our measurements, accuracy tells how correct those reading actually are. Both concepts are important, but it's critical to understand that one does not necessarily imply the other. For instance, our system may be capable of resolving down to 0.5°F, but that may not be of much use if these highly specific measurements are off by 10°F because of problems in the system.

Where might such egregious errors arise? The first place to look is in the sensing element itself, since an inaccurate sensor will introduce imprecision that may or may not be repeatable (the former being more desirable since it can be compensated). Indeed, different types of thermocouples have differing degrees of accuracy over their operational range, with J-type thermocouples being the most accurate (±0.1°C); E, R, and T thermocouples good to ±0.5°C; K thermocouples accurate to ±0.7°C; and S thermocouples accurate to ±1.0°C.

Other problems in the signal chain can cause incorrect results as well, including a lack of or insufficient cold-junction compensation (discussed in the section titled Cold-junction Compensation), failure to ensure proper common-mode noise rejection, or the introduction of non-common-mode noise into the circuit. Usually, we can compensate for inaccuracies due to the sensor itself or due to other elements in the circuit, provided those problems are not time- or temperature-dependent, through proper calibration and linearization. The key points to remember are that resolution is not equivalent to accuracy and that we need to have both in order to have a robust, reliable system.

Challenges

We now shift our focus from the general to the particular, from the aspects of system design that apply to all sensor systems to those that we face specifically because we're using thermocouples as our sensing element.

Signal Characteristics

The starting point for any sensor system is to determine the characteristics of the signal output by the sensor over the range of interest. These characteristics include the voltage levels of the signal, the sensor's output drive capabilities, and the anticipated frequency content of the signal.

Signal Level

As a quick check of the NIST thermoelectric tables shows, thermocouples produce a very small output voltage, on the order of a few millivolts. Complicating our use of the thermocouple signal is the fact that it's essentially a zero-current signal, meaning that it's only able to drive very high-impedance loads. Unfortunately, the dsPIC's ADC input impedance of about 20 KΩ doesn't meet that criterion. Finally, depending upon the temperature range that we want to measure, the thermocouple's output voltage can be negative, which violates the input signal requirements for the dsPIC's ADC. Even if the thermocouple normally produces a positive voltage for all of the temperatures of interest, should the thermocouple's leads become reversed (a condition that occurs in practice more frequently than one might imagine), the input voltage to the digitization circuitry still may be negative. Clearly, a thermocouple-based system requires a high-impedance amplifier between the sensor output and the ADC input to buffer, amplify, and level-shift the signal.

By choosing the amplifier's gain and offset appropriately, we can design a system that works when the thermocouple is wired correctly or when it is reversed. At first blush, this capability may not seem to be particularly important; after all, couldn't the user simply swap the thermocouple wires to correct the problem? Sometimes this is a viable option, but there are circumstances in which it is not. On one injection-molding project with which the author was involved, the cost to remove the mold (where the thermocouples were installed) and to make any change to it started at $50,000, and the cost to run the machine was $20,000 *per day*. The complete system supported over a hundred sensors of various types, and Murphy's Law[11] dictated that at least one of the thermocouples would be miswired (which actually happened). In this environment, the ability to operate with minor wiring problems saved a great deal of money and aggravation for the customer.

Frequency Content

If we intend to filter the sensor inputs to reduce electrical noise, we need to identify the frequency content of the sensor signal so that we know the frequencies that we can safely attenuate without excessively degrading the sensor signal itself. To determine the spectral content of the temperature signal, we need to understand something of the thermal characteristics of the system that we're monitoring, namely how quickly the temperature can change over time. Some systems can change temperature very quickly, while others may fluctuate relatively slowly. Although the designer frequently doesn't have a hard specification for this in advance, he often has a good feel for the maximum rate of temperature change that can be expected. As an example, in a heating application, based on the maximum power we can apply to the heating element and the thermal dissipation properties of medium being heated, we should have a good idea of the maximum expected rate of temperature change. When we turn on an oven to heat a pizza, we don't expect it to instantly heat to 400°F; it should take maybe three minutes. Using this admittedly crude approach, we would estimate that our system has a maximum temperature velocity[12] of approximately:

$$\Delta T_{MAX} = 400 \text{ °F} / 180 \text{ seconds} = 2.22 \text{ °F/sec}$$

If we continue with the 12-bit J thermocouple example that we developed in the Resolution of Measurement section, and remembering that 1°C = 1.8°F, we can see that this maximum rate of temperature change corresponds to an ADC count of:

$$\Delta Count_{MAX} = \Delta T_{MAX} / T_{RES}$$

$$\Delta Count_{MAX} = (2.22°F/sec) / (0.2124°C/count * (1.8°F /°C))$$

$$\Delta Count_{MAX} = 5.8 \text{ counts/sec} \approx 6 \text{ counts/sec}$$

Assuming that we want to be able to track temperature changes of only 1 ADC count, this would imply that we need to sample at least 6 times per second. However, applying the Nyquist criteria that we learned in Chapter 2, the theoretical minimum sampling rate should be at least twice that (12 samples per second) and the practical sampling rate should be around five times that (30 samples per second).

There are all sorts of caveats to using this approach, starting with the most obvious one: there may be, and probably are, periods during the heating cycle in which the temperature changes by more than the linear amount predicted using this technique. In addition, we're assuming that the appropriate analog antialiasing filters have been applied to the sensor signal to suppress broader-band noise; if that is not the case or if the filter bandwidth is broader than required by the sampling rate, we need to increase the sampling rate appropriately. Finally, thermocouples themselves have response-time characteristics, meaning that it takes a finite amount of time for the thermocouple output voltage to accurately reflect the temperature at its junction. Nonetheless, in the absence of any additional information, the technique is probably a good starting point.

Cold-junction Compensation

By far the most difficult and least understood aspect of interfacing with thermocouples is a concept known as *cold-junction compensation* or CJC. As you'll recall from the earlier discussion of thermocouples, the Seebeck effect generates a temperature-dependent voltage at the junction of two dissimilar metals, and it's this property that makes thermocouples so useful to us. Unfortunately, the Seebeck effect is not limited to the junction of the two thermocouple leads with each other (known as the "hot" junction); it also occurs at the termination of the thermocouple leads (known as the "cold" junction) into the copper traces on the circuit board to which they are connected. It is this unwanted Seebeck-effect voltage that we have to remove, and we do so using a technique known as *cold-junction compensation.*

The basic idea behind cold-junction compensation is to measure the temperature at the cold junction (i.e., where the thermocouple leads enter the PCB) and to add that temperature to the temperature calculated for the hot junction (i.e., at the sensor's thermocouple), thus compensating for the unwanted Seebeck-effect voltage.[13] An equivalent approach is to simply add the corresponding cold-junction voltage to that for the hot junction. To do this properly, the design must ensure:

1. that the terminations at the cold junction for both thermocouple leads are at the same temperature (known as an *isothermal termination* or *isothermal barrier*), and

2. that the device used to either measure the temperature of the isothermal barrier or to generate a corresponding voltage is located as closely as possible to the isothermal barrier.

Failure to meet both requirements will introduce an uncorrectable error into the temperature measurements and, even worse, this error will change with temperature.

Linearization

Next to cold-junction compensation, the most difficult task when processing thermocouple signals is the mapping of the nonlinear thermocouple voltage to a linear temperature scale. One common way of doing this is by evaluating a high-order polynomial expression using a recursive technique with thermocouple type-dependent coefficients, but since it is a mathematically intense operation, this approach is generally unsuitable for embedded-processor applications. Although the dsPIC DSC can perform mathematical operations very efficiently, there's another technique known as *piecewise linearization* that yields excellent results while being less complex to implement.

The basic idea in piecewise linearization is to divide a nonlinear curve into a series of linear segments, as shown in Figure 6.2. Within each segment, we approximate the value of the actual curve by the value of the line between the segment's starting point and ending point. To compute the estimated value, we simply determine the segment to use by finding the one containing the value we're mapping, and then calculate the estimated value using the slope and Y-intercept of the corresponding segment. We thus reduce a high-order polynomial computation to a search and a linear computation.

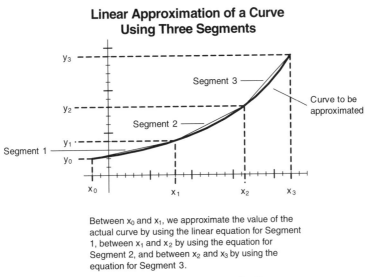

Between x_0 and x_1, we approximate the value of the actual curve by using the linear equation for Segment 1, between x_1 and x_2 by using the equation for Segment 2, and between x_2 and x_3 by using the equation for Segment 3.

Figure 6.2. Example of Piecewise Linearization of a Curve

Two things determine the technique's efficacy: the number of segments used to parse the curve and the nature of the curve we're attempting to approximate. Obviously, the more segments we use, the more accurately we can match the curve, but also the more time required to identify the appropriate segment to use in the linearization. This balancing act thus becomes one more tool in the designer's kit, allowing him to trade accuracy for speed or vice versa.

Calibration

Linearization is one important part of converting the temperature voltage signal into a value that can be used by the dsPIC firmware, but it is not the only process required to generate a useful digitized value. The linearized signal must also be *calibrated* to a set of known temperature readings to ensure that the computed signal value is accurate for the particular hardware. By far the most widely used technique for calibrating thermocouples is the *two-point method*, in which two readings are taken at different temperatures (usually at points near either end of the temperature range of interest), and those points are then used to create a linear calibration reference curve.

As an example, suppose we want to measure temperatures between 100°F and 900°F. We might take as our calibration points the temperatures 150°F and 750°F, since they are near the endpoints of our range and ensure that the linear curve we generate is not too far off of the actual thermocouple curve. Although we could use 100°F and 900°F as our endpoints, this would result in a linear curve that has a greater maximum error, as shown in Figures 6.3a and 6.3b.

Calibration Curve with 150 ºF and 750 ºF as Calibration Points

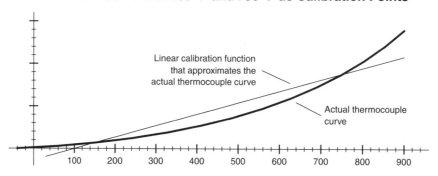

Note that the bow of the actual thermocouple curve is exaggerated to better illustrate the error between the linear calibration function and the actual thermocouple curve. With this choice of calibration points, there is less error between the two calibration points, but the error increases outside of the two calibration points.

Figure 6.3a. Calibration Curve Using 150°F and 750°F

Calibration Curve with 100 ºF and 900 ºF as Calibration Points

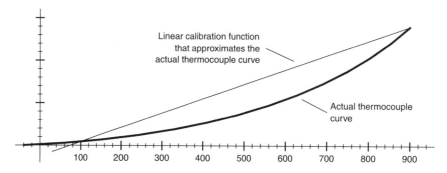

Note that the bow of the actual thermocouple curve is exaggerated to better illustrate the error between the linear calibration function and the actual thermocouple curve. This choice of calibration points creates a significantly larger error between the two calibration points but reduces the error at the very top end of the measurement scale (near 900 ºF.)

Figure 6.3b. Calibration Curve Using 100°F and 900°F

To perform the calibration, we would measure the linearized voltage at the lower calibration point, repeat the measurement at the upper calibration point, and calculate the gain of the calibration curve as:

$$G_{CAL} = (T_{UCP} - T_{LCP}) / (V_{UCP} - V_{LCP})$$

$$G_{CAL} = (750°F - 150°F) / (V_{UCP} - V_{LCP})$$

$$G_{CAL} = 600°F / (V_{UCP} - V_{LCP})$$

While the offset is simply:

$$OFF_{CAL} = T_{LCP}$$

In these equations,

G_{CAL}	is the gain for the calibrated reference curve
OFF_{CAL}	is the offset for the calibrated reference curve
T_{UCP}	is the temperature of the upper calibration point (750°F)
T_{LCP}	is the temperature of the lower calibration point (150°F)
V_{UCP}	is the voltage reading at the upper calibration point
V_{LCP}	is the voltage reading at the lower calibration point

In practice, the sensor firmware would take a voltage reading, linearize it, and then convert it to the corresponding temperature through the equation:

$$T_{CAL} = G_{CAL} \times (V_{READING} - V_{LCP}) + OFF_{CAL}$$

where

T_{CAL}	is the calibrated temperature reading
G_{CAL}	is the gain for the calibrated reference curve
OFF_{CAL}	is the offset for the calibrated reference curve
$V_{READING}$	is the linearized voltage reading
V_{LCP}	is the voltage reading at the lower calibration point

All of this discussion raises one fairly obvious question: if we're trying to calibrate the temperature sensor, how do we "know" what the temperature is at our two calibration points? The answer is that we have to use a calibrated source, typically a thermocouple calibrator that injects a precision voltage based on the specified temperature. These devices are available at relatively low cost from a number of suppliers, although the more accurate and easier to use versions can cost several hundred to several thousand dollars.

Sources of Noise

That there will be sources of noise corrupting the thermocouple system is a given and, with its the thermocouple's intrinsically low signal level, the effects of electrical noise can be devastating. The good news is that we can use both the inherent noise-cancelling characteristics of the thermocouple's differential signal and the frequency spectra of the temperature signal and of the noise to effectively filter out a significant portion of the unwanted signal.

AC Power

A common source of electrical noise arises from AC power-line radiation, particularly in industrial or other environments in which there are cables carrying large currents near the thermocouple. Power-line noise does have one redeeming quality: its spectral content is extremely narrow-band and is centered about either 50 Hz or 60 Hz (depending on location), with harmonics at multiples of those frequencies. To a designer, such well-defined narrow-band noise signals are actually easier to handle than lower-level broadband noise, provided that the bandwidth of the noise signal does not encompass too significant a portion of the frequency band of the sensor signal itself.

Because the thermocouple signal is so small and drives virtually no current, it's extremely susceptible to power-line contamination. Fortunately, thermocouple signals do have two things working in their favor: their differential nature helps reject common-mode noise and the frequency spectrum of the power signal is often outside of the temperature signal band. Even in those cases in which the power-signal frequencies are within those of the temperature signal, the narrow-band nature of the power signal usually allows the system to remove it effectively using a band-stop filter.

Figure 6.4a shows an example of an out-of-band power-line signal's frequency content along with a spectrum for a sample temperature signal. Using the low-pass filter shown in Figure 6.4b, we can easily remove the spurious spectral content, producing the resultant filtered signal shown in Figure 6.4c.

The more difficult case of in-band power-line noise is shown in Figure 6.5a, in which the power-line signal is within the spectra of the temperature signal. In this case, we have to employ a high-order notch filter to surgically remove the unwanted frequencies, which leaves an improved, but clearly not perfect, filtered temperature signal. Even with this slight degradation, the resultant signal is more accurate than the unfiltered version.

Temperature Signal with Out-of-Band Power-line Noise

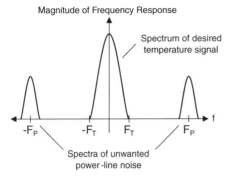

Figure 6.4a. Sample Temperature Signal with Out-of-band Power-line Noise

Low-pass Filter to Remove Power-line Noise

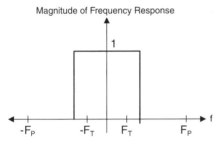

Figure 6.4b. Low-pass Filter to Remove Out-of-band Power-line Noise

Filtered Temperature Signal with Power-line Noise Removed

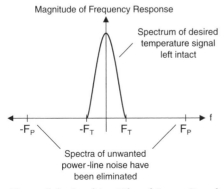

Figure 6.4c. Resulting Filtered Sensor Signal

Temperature Signal with In-Band Power-line Noise

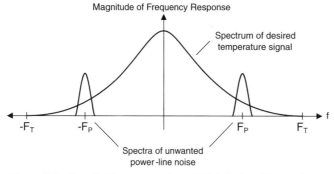

Figure 6.5a. Sample Temperature Signal With In-band Power-line Noise

Notch Filter to Remove In-Band Power-line Noise

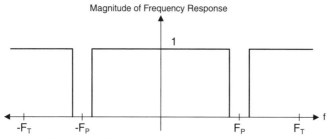

Figure 6.5b. Notch Filter to Remove In-band Power-line Noise

Filtered Temperature Signal with Power-line Noise Removed

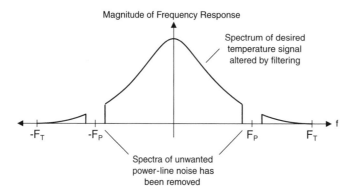

Although the spectrum of the desired temperature signal has been affected by the filtering, most of the original signals is intact so we should be able to obtain useful results.

Figure 6.5c. Resulting Filtered Sensor Signal

Finally, Figure 6.6 illustrates the case where the power-line noise spectrum significantly overlaps that of the temperature signal itself. In this case, there is no good filtering approach that will compensate for the power-line noise, and the designer is left with the unsatisfactory alternative of trying to correlate the power-line noise on the thermocouple with the power-line signal itself and then subtracting this from the measured sensor signal. While technically possible, a much better approach for such a situation is to shield the thermocouple and the power-line cables and to route them so as to minimize the coupled radiation. In practice, it is exceptionally unlikely that most thermocouple-based systems will ever encounter such an extreme circumstance since there are very few materials (including those for thermocouples) that can respond thermally at a 50-Hz or 60-Hz rate.

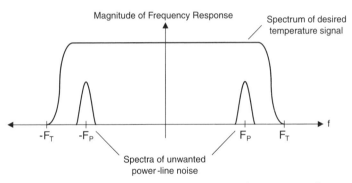

Figure 6.6. Sample Temperature Signal with Overlapping Power-line Noise

Error Conditions

There are two serious error conditions that can be detected fairly easily through the judicious use of high-impedance biasing resistors attached to the thermocouple and by paying attention to the characteristics of the measured thermocouple voltage. Failure to check for these conditions can cause the sensor to report grossly inaccurate measurements and, depending upon how these measurements are being used can lead to catastrophic system failure.

Open Thermocouple

The first error condition we can check is the presence of an open thermocouple—i.e., a thermocouple in which the junction between the two thermocouple leads has been broken. This can occur fairly frequently in systems that experience severe mechanical or thermal stress, particularly if that stress is repeated over time. Detection is straightforward provided that two matched high-impedance resistors are added to the thermocouple input connection (one per lead) as shown in Figure 6.7. Should the junction break, the leads will be biased to the extreme of the input voltage range, a condition that is easily detected by the dsPIC DSC.

**Sense Resistors Added to
Thermocouple
Inputs to Detect Error Conditions**

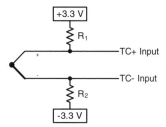

Figure 6.7. Sense Resistors Added to Thermocouple Inputs

In selecting the bias resistor value, the designer does not want to swamp the signal produced by the thermocouple itself under normal circumstances. Something on the order of 100 kΩ generally works quite well, but it's important to match the two resistors as closely as possible to maintain the differential nature of the signal.

Reversed Thermocouple

Since the thermocouple leads are polarized, it's possible to reverse them. Although this won't damage the input circuitry because the signal is of such a low level, if undetected, the condition will read erroneous results. When a thermocouple is reversed, the measured signal voltage decreases as the temperature increases, but since we don't usually know whether the temperature is actually increasing or decreasing, the only way to be confident that the thermocouple is reversed is to detect a voltage that is below the expected lower range of the signal. For instance, a J thermocouple produces negative voltages when exposed to temperatures below 0°C, so if the range of interest is between 0°C and 500°C (32°F and 932°F), then any negative voltages would indicate that the thermocouples were reversed.

Because we want to be able to discern when the thermocouple output drops below a certain minimum value and since we want to be able to determine when the thermocouple itself is open, it's important to allow a little headroom at either end of the digitization range. By doing so, we can assure that we measure the entire temperature range of interest while still being able to correctly identify faults with the thermocouple.

6.3   Application Design

With all of that background information under our belt, we're now ready to design a real-world intelligent temperature sensor (at least as real-world as one can get using a demo board as the hardware platform). Although a basic design, the application described here can be extended easily to perform more complex tasks, such as temperature control as well as sensing.

As with any application, the first step is to specify the required system functionality, which we'll do in the section titled System Specification. With the system specs in hand, the next steps are to design the sensor-specific signal conditioning circuitry, to determine the digital-filtering requirements, and to define the data-analysis algorithms that need to be performed, all of which sections Sensor Signal Conditioning, Digital Filtering Analysis, and Data Analysis Algorithms cover. Finally, the sensor has to communicate its results to the rest of the system using a predefined protocol, which is discussed in the section Communication Protocol.

System Specification

Our goal is to develop a generic temperature sensor hardware and software platform, so the specifications are not particularly detailed because we don't want to get bogged down in minutiae that are applicable only to a few select situations. Be aware, though, that often the specifications for a sensor system are extremely detailed, and the engineer needs to review them carefully prior to performing any detailed design to ensure that they are achievable.

The temperature sensor developed here will meet the following functional requirements:

1. Sample one thermocouple and one cold-junction compensation channel.

2. Sample each channel 500 times per second, assuming a maximum temperature signal frequency content of 15 Hz (approximately 33× oversampling).

3. Support J- and K-type thermocouples.

4. Perform required cold-junction compensation for both types of thermocouples.

5. Screen for open and reversed thermocouple conditions.

6. Filter the sampled data to remove power-line noise.

7. Allow the user to perform the following functions via the RS-232 serial port running at 38.4 Kbps, 8 data bits, 1 stop bit, no parity, and no flow control:

a. perform calibration of each channel, and

b. specify a lower and an upper temperature alarm limit for each channel.

8. Report the measured temperatures for all channels every second via the serial port. If the system detects an error condition for a particular thermocouple, that error condition will be reported instead of the temperature reading. Temperature and error reports are to be in text.

9. Report out-of-limit alarm conditions by displaying a sensor value of "----" and lighting LED1 on the demo board if the measured temperature is less than the lower limit and by displaying "++++" and lighting LED2 if the temperature exceeds the upper alarm limit.

This application monitors only two signal channels to reduce the amount of circuitry that the reader has to implement; there is nothing from a performance standpoint that would prevent the system from monitoring any number of channels up to the full 16 supported by the dsPIC30F6014A chip. While the sample rate may seem a bit high in absolute terms, it is one that allows us to easily remove power-line noise without significant signal delay through the system. As with the number of supported channels, the sampling rate can be increased if necessary; the dsPIC DSC has plenty of processing power to do so. Also note that the reporting rate of the measured temperatures is so low because the reported data is intended to be read by humans. If the data is to be read by other electronic components in the system, a much faster binary-oriented protocol would be appropriate, code for which is also supplied in the book's CD.

Sensor Signal Conditioning

As described earlier in the chapter, the thermocouple sensor interface needs to

1. buffer and amplify the thermocouple output signal,

2. perform cold-junction compensation, and

3. detect sensor error conditions.

In addition, we also need to include an analog antialiasing filter that effectively eliminates frequencies above 250 Hz (since the sampling rate of 500 Hz has to be at least twice the highest frequency content of the temperature signal). Since the temperature signal that we're sampling is assumed to have a maximum bandwidth of 15 Hz (meaning that we're sampling 33 times faster than is theoretically necessary), we should use this information to further tighten the antialiasing filter cutoff frequency to 15 Hz as well. The circuitry shown in Figure 6.8 does all of this for a single channel. Let's examine each portion of the circuitry to gain a better understanding of what's included and why. "Note that the schematic is configured for operation at 3.3V. If the circuit is implemented on the dsPICDEM 1.1 GPDB (which uses 5V supplies), change all references to 3.3V to 5V, all –3.3V references to –5V, and change 1.65V to 2.5V.

Thermocouple Interface Circuitry

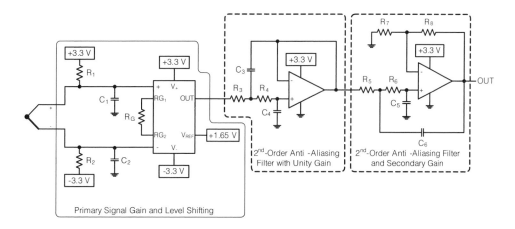

NOTES:
1. R_1 and R_2 provide both an input common-mode current path and a way to determine whether the thermocouple is actually connected to the circuit Resistor values should be identical and be between 10 KΩ and approximately 100 KΩ to avoid loading the input signal with too low an input impedance while still providing an input common-mode current path that has a low enough impedance to be effective.

2. Gain through the instrumentation amplifier is controlled by R_G according to the equation
$$\text{Gain} = 1 + (50\ K\Omega\ /\ R_G)$$

3. C_1 and C_2 remove high-frequency input noise. Capacitors are placed on both input signals to maintain a balanced differential input impedance.

4. Anti-aliasing filters have a Butterworth filter frequency response. Gain is included in the final filter stage to compensate for the instrumentation amplifier's inability to produce output voltages that are near either power supply rail. The gain in the final stage is given by the equation
$$\text{Gain} = 1 + (R_8\ /\ R_7)$$

Figure 6.8. Schematic of Thermocouple Interface (Single Channel)

Differential Amplifier

The INA326 instrumentation amplifier provides buffering and amplification of the differential thermocouple signal. Instrumentation amplifiers are perfect for this type of application since they have very high input impedance and are inherently differential. Note the biasing resistors that are connected to each side of the thermocouple signal to provide open thermocouple detection and that provide a DC bias path required by the instrumentation amplifier for proper operation.

Antialiasing Filter

The antialiasing filter has one primary task: to keep frequencies higher than half the sampling frequency from contaminating the sampled signal. At first glance, it might seem that we would want to use a high-order analog filter for this task so that we could get extremely strong filtering, but in many cases this approach is not the best. In a previous chapter, we've alluded to the shortcomings associated with analog filters (drift, cost, board space), and these problems are exacerbated the more complex the filter becomes. Instead, we'll go with a pair of cascaded second-order Butterworth analog filter up front and then do the heavy lifting with our DSP filters. This filter structure is optimally flat through the passband and has a fairly sharp rolloff of 20 dB/decade/pole,[14] as can be seen in Figure 6.9.

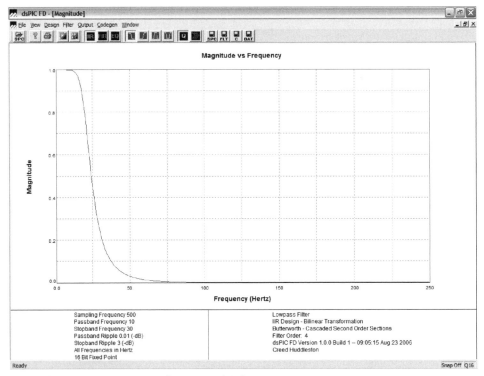

Figure 6.9. Frequency Response of Butterworth Filter

Digital Filtering Analysis

In this example, the power signal is considered to be out-of-band noise, allowing us to use a sharp low-pass digital filter to reduce noise on the temperature signal. If our temperature signal were more broadband, we would need to include a notch filter also, as we'll see in the next chapter.

There are a number of packages available with which to design digital filters, but the one used for this book is the Microchip Digital Filter Design System, v1.0.0. As with the other development software discussed here, the filter design package is available through Microchip.[15] One of the nice aspects of using a software package to design the filter is that it's easy and quick to get graphical feedback on the filter's characteristics. For instance, Figures 6.10 and 6.11 show the effects of specifying a looser (Figure 6.10) and a tighter (Figure 6.11) stopband ripple requirement. As we can see, tightening the ripple specification also affects the passband ripple and the width of the passband window, introducing more ripple into the passband while both broadening the passband a bit and creating a sharper transition between the passband and the stopband.

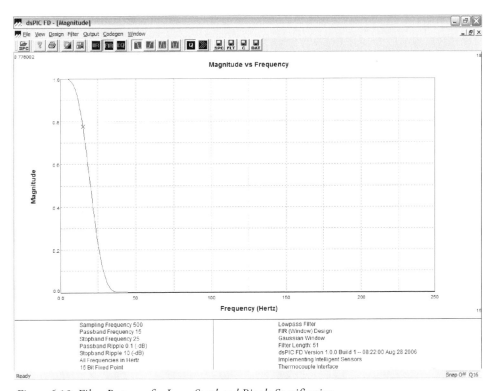

Figure 6.10. Filter Response for Loose Stopband Ripple Specification

Although the filter with the looser stopband ripple requirement initially might look like a better filter since its passband is flatter, as we can see from the upper left corner of its Magnitude vs. Frequency graph in Figure 6.10, the filter passes just over 78% of the input signal at 6 Hz. In contrast, while the filter with the tighter stopband ripple requirements obviously has more ripple in both the passband and the stopband (not what we'd necessarily expect, but still within spec), it also passes more of the signal at 6 Hz (82%) and has a sharper transition between the passband and the stopband. All of this demonstrates that filter design is a game of compromise, with the designer having (or getting, depending on one's perspective) to trade off better performance in one area for worse performance in another.

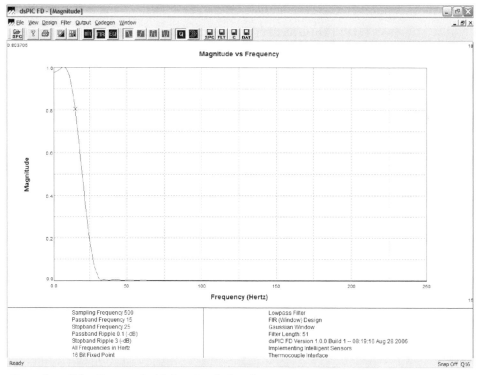

Figure 6.11. Filter Response for Tighter Ripple Specification

For this application, we'll go with the somewhat looser filtering requirements since they meet our needs and don't affect the signal quite as adversely. The principle of filtering as lightly as necessary (but still enough to get the job done) is a sound one to follow in general, since it avoids excessive modification of the original signal. In this regard, the principle is not too far from the injunction to physicians to "first, do no harm."[16]

One other aspect of the digital filter should be noted. In both cases, the filter has 51 taps. While we could certainly improve the performance of the filter by adding more taps, doing so would add to the delay the signal experiences going through the filter. For this application, we intentionally selected a relatively short filter to allow the system to be more responsive.

Data Analysis Algorithms

The data analysis needs for this application are pretty simple, basically requiring that we check each filtered temperature sample against a lower and an upper alarm limit that's been set by the user, and lighting an alarm whenever one or the other has been exceeded. A flow chart illustrating this rather limited data analysis is shown in

Figure 6.12. Although the analysis presented here is admittedly simple, notice how easy it would be to extend it to include more complex analysis or perhaps to add an output-control algorithm. This is the type of structure that designers should try to employ whenever possible: simple, clean, and easily extensible. Not only is such a framework more reliable, it also is easier to add in new features as the product matures.

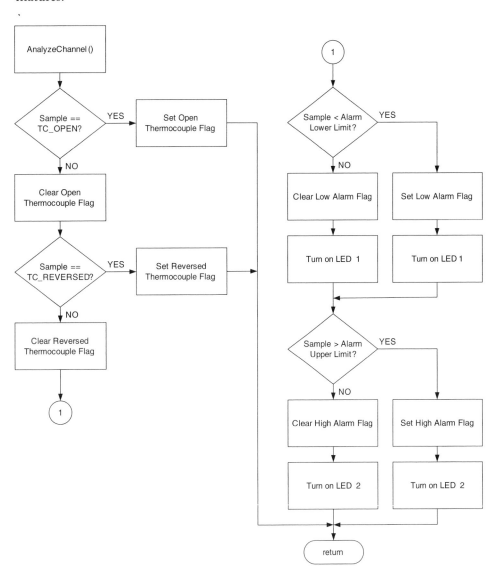

Figure 6.12. Data Analysis Flow Chart for Thermocouple Sensor

Communication Protocol

Application developers sometimes mistakenly equate a communication interface that works under nominal conditions with one that works under all conditions, including errors in the physical medium used by the communication interface. A good designer takes into account all situations that are likely to arise and develops a way for the system to degrade gracefully when it encounters unanticipated conditions as well. This "graceful degradation" is extremely important; even if the system can't correct an error, it should alert its host and then take any actions required to ensure that it is in a state in which the system can operate safely. Implementing a state-based communication interface that automatically resets itself whenever it encounters an unexpected input goes a long way toward that goal.

This first application will use a simple (there's a theme here), human-readable protocol over a standard RS-232 serial port. Not only does this approach allow the user to actually interact with the system, it also allows us to wade, not plunge, into the world of state-based communication handlers. It's the latter aspect that's the more important of the two, because a surprisingly large number of products are deployed with weak, unreliable communication interface implementation. There are few things more frustrating from a user's perspective than to experience unexplained system failures (often with the appearance that the system has "locked up"). When such failures arise because of poorly implemented communications, they are simply inexcusable.

Having extolled the benefits of state-based communication handling, let's now take a closer look at what we mean by that term and how we design using state-based approach. In a state-based handler, the actions that the application takes are dependent upon both the current state of the handler (a value that is stored in a global variable) and the data value being processed. Depending upon the action taken, the state value may change or it may stay the same. An example should help clarify the concept.

Figure 6.13 shows a state diagram for a simple communication handler. Each circle in the diagram represents an individual processing state, and the lines extending from a particular processing state are *transition conditions*, the conditions required to change from one state to another. If no condition is given for a particular transition line, then the state machine automatically moves from the originating state to the terminating state once all of the processing for the originating state has been performed. For instance, looking at the *Initialize* state, we see one transition line extending from it to the *Accept Input* state, and that transition line has no associated conditions, meaning that the state machine will cycle from the *Initialize* state

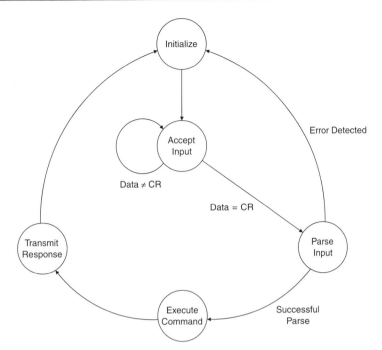

Figure 6.13. Simple Communication Handler State Machine

to the *Accept Input* state as soon as all of the processing for the *Initialize* state has completed.

Once the state machine enters the *Accept Input* state, it remains there until it receives an ASCII carriage return character as input. This is indicated on the diagram by two transition lines, one with a transition condition of *Data ≠ CR* that simply loops back to the *Accept Input* state and the second with a transition condition of *Data = CR* that points to the *Parse Input* state. Whenever the state machine receives new data, it processes it in the *Accept Input* state and then either stays in that state (if the data is not a carriage return) or transitions to the *Parse Input* state (if the data is a carriage return). Typically, the processing performed in the *Accept Input* state would be something along the lines of adding the new data to a buffer, assuming that there is room in the buffer, although it could be more sophisticated than that if required by the specific application. In any event, receipt of a carriage return signals the state machine that it should parse the received data to identify the action that the user would like the system to perform.

If the *Parse Input* processing is able to successfully extract the user's command from the input data buffer, the state machine transitions to the *Execute Command* state, which performs the task specified by the user's command input. If, on the other hand, the parsing detects an error in the command input, it will cycle the state machine back to the *Initialize* state, resetting the state machine to a known good state to await the user's next input. By handling the processing like this, the system's operation is insulated from errors in the user's input. Users can (and probably will) make errors, but those errors will not affect the safe operation of the system.

The processing performed in the *Execute Command* state simply performs the task associated with the specified command and command parameters. Once the task has completed, the state machine moves to the next state, *Transmit Response*, to report the results of the command execution. Finally, after transmitting the command response back to the host, the state machine cycles to the *Initialize* state to reinitialize the machine's state variables in preparation for reception of the next command.

Although it's certainly possible to implement the state machine in hardware, for example in a programmable logic chip, it's far more convenient, flexible, and less expensive to implement the state machine in software (assuming, of course, that the processor can cycle through the states quickly enough to meet all timing requirements). Doing so is straightforward, but some guidance may be helpful for those who have never designed a software state machine.

When implementing a software state machine, the application uses global *state variables* to track both the specific operational state of the machine and any information that has to be preserved when cycling to another state. The state machine itself, at least at the highest levels, is usually implemented as a single function that is called repeatedly with new input data to cycle the machine through the desired sequence. If the application is using the C language, this function often takes the form of a `switch` statement, with the parameter for the statement being the current processing state. Using the example that we've been developing, the code would look something like the following.

Code Example 6.1. Implementation of Simple Communication State Machine

```
                    // Communication State Definitions
#define COMM_ST_INIT            0       // Initialize state machine
#define COMM_ST_ACCEPT_INPUT    1       // Accept input data
#define COMM_ST_PARSE_INPUT     2       // Parse complete line of input
data
#define COMM_ST_EXEC_CMD        3       // Execute the parsed command
#define COMM_ST_TX_RESPONSE     4       // Transmit the command response

                    // State Variables
```

```
Uint8
    g_ui8CommState = COMM_ST_INIT;  // Global communication processing state


/**********************************************************************
*   FUNCTION:      CommProcRxData(Uint8 ui8NewData)                   *
*                                                                     *
*   DESCRIPTION:   This function implements the high-level software   *
*                  state machine for communication processing. It is  *
*                  called by the main processing loop whenever the    *
*                  processor receives new input data, and the function *
*                  cycles through the appropriate states based on the  *
*                  current processing state and the new input data.   *
*                                                                     *
*   PARAMETERS:    ui8NewData - new input data to process             *
*                                                                     *
*   RETURNS:       The function returns the new communication processing *
*                  state.                                             *
*                                                                     *
*   REVISION: 0    v1.0              DATE: 12 June 2006               *
*       Original release.                                            *
**********************************************************************/


Uint8
    CommProcRxData(Uint8 ui8NewData)
    {
    // Local Variables
    Bool
        bProcessingState = TRUE;    // Continue processing state flag (set to
                                    //   continue processing states, cleared
                                    //   to stop processing after the current
                                    //   state)
    // Identify the current communication state
    // and process the new data accordingly
    while (bProcessingState)
        {
        switch (g_ui8CommState)
            {
            case COMM_ST_INIT:
                // Initialize the comm state machine
                ...                         // Perform state-specific processing
                bProcessingState = FALSE;   // Finished processing for this cycle
                // Cycle to next state
                g_ui8CommState = COMM_ST_ACCEPT_INPUT;
```

```
            break;
    case COMM_ST_ACCEPT_INPUT:
        //  Accepting new input data
        …                       // Perform state-specific processing
        //  Cycle to next state
        if (ui8NewData == '\r')
            g_ui8CommState = COMM_ST_PARSE_INPUT; //  Parse the
                                                  //     data if
                                                  //     we've
                                                  //     received
                                                  //     the entire
                                                  //     line
        else
            bProcessingState = FALSE;    // Finished processing for this
                                         //    cycle
        break;
    case COMM_ST_PARSE_INPUT:
        //  Parse the complete input data buffer
        …              // Perform state-specific processing that
                       //    returns a value of bError to indicate
                       //    whether the parsing was successful

        //  Cycle to next state based on
        //  the success of the parsing
        if (bError)
            g_ui8CommState = COMM_ST_INIT; // Error detected so
                                           //     reset
                                           //    the state machine and
                                           //    wait for next command

        else
            g_ui8CommState = COMM_ST_EXEC_CMD;    // No error so
                                                  //     execute the
                                                  //     command
        break;
    case COMM_ST_EXEC_CMD:
        //  Execute the parsed command
        …                       // Perform state-specific processing
        //  Cycle to next state
        g_ui8CommState = COMM_ST_TX_RESPONSE;
        break;
    case COMM_ST_TX_RESPONSE:
        //  Transmit the command response
```

```
        ...                        // Perform state-specific processing
        //  Cycle to next state
        g_ui8CommState = COMM_ST_INIT;
        break;
      default:
        //  Should NEVER get here so reset
        //  the state machine
        g_ui8CommState = COMM_ST_INIT;       // Cycle to next state
        break;
    }
  }
  return g_ui8CommState;
}
```

There are a number of points to note about the software state-machine code. In it, we use a global variable (g_ui8CommState) to maintain the processing state of the machine. The code initializes the value of g_ui8CommState in the variable's declaration to ensure that the state machine starts in a known state (the initialization state), and the function updates the value of g_ui8CommState whenever the state machine needs to change to a new state. The function returns the final state of the machine after the routine has performed all processing for the current state, but since the current state value is stored in a global variable already, the return value is really for informational purposes only. To ensure proper operation, only CommProcRxData() should modify g_ui8CommState. Even if other routines monitor the state by examining the value returned by CommProcRxData(), they should not set its value because this can cause deeply hidden software interactions that may cause problems should the state-machine implementation change in future releases.

The state diagram in Figure 6.13 shows that some states require input or the occurrence of a specific condition in order to cycle to the next state. As structured here, the function cycles between states that don't require any user input in a single call to CommProcRxData(). Depending upon the amount of processing that must be done and the length of time that processing takes, this approach might need to be modified.

6.4   Hardware Implementation

The hardware block diagram shown in Figure 6.14 shows the major hardware building blocks that will be discussed in the following sections. Note that this is one way of designing the circuitry, not the only way, and those with greater analog design skills should feel free to improve upon the circuitry presented here.

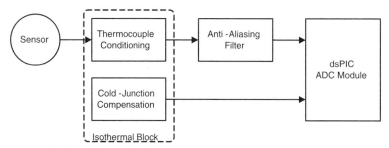

Figure 6.14. Hardware Block Diagram

Analog Amplifier and Antialiasing Filter

A schematic for the analog amplifier and anti-aliasing filter sections is shown in Figure 6.8. In it, we see that the amplifier has been implemented using a single-chip instrumentation amplifier, with the output of the instrumentation amp feeding into a fourth-order Butterworth low-pass antialiasing filter constructed of two rail-to-rail op-amps and a few passive components. There are three key design considerations in this section:

1. minimization of power consumption to reduce self-heating effects,

2. clean, compact printed circuit board layout to minimize crosstalk and noise, and

3. good power-supply bypassing to reduce noise.

Low power consumption is achieved primarily through the use of low-power ICs, a good many of which are available from a variety of suppliers. Because thermocouple outputs are low-drive and low-level signals, the circuit layout is critical. Traces to the thermocouple terminations should be as short as possible, and those for a given thermocouple should be side-by-side and close together, if possible, to improve common-mode immunity (this ensures that thermal and electrical noise is coupled onto both leads equally). Finally, using a good solid power supply with minimal ripple and drift is extremely important, or noise from the power supply itself (as opposed to noise on the thermocouple output) will be introduced into the system, too.

Cold-junction Compensation

While there are a number of ways to implement cold-junction compensation, the circuitry here uses a cold-junction compensation chip, the Linear Technologies LT-1025, to perform that function in hardware as shown in the schematic in Figure 6.15. Selection of either J or K thermocouple compensation is hardware selectable by moving the position of jumper J1.

Cold-Junction Compensation Interface

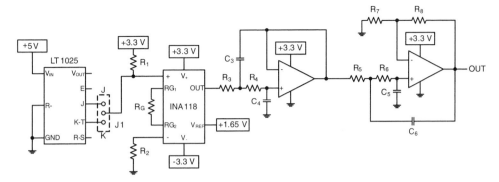

NOTES:
1. Resistor and capacitor values should match those used for the thermocouple signal processing channels.

2. Jumper J1 should be set to connect the "J" pin to the instrumentation amplifier if compensating for a J thermocouple or set to connect the "K" pin to the instrumentation amplifier when compensating for a K thermocouple.

Figure 6.15. Cold-Junction Compensation Schematic

Here the LT1025's voltage output is buffered and then measured by the dsPIC DSC, which then uses a software algorithm to remove the corresponding cold-junction voltage from that measured for the thermocouples. For the greatest accuracy, one LT1025 can be used per thermocouple input, but in practice it is usually fine to share a single LT1025 for two thermocouples if the LT1025 is placed between the pairs for each thermocouple (i.e., locating the pair of leads for one thermocouple on one side of the LT1025 and the leads for the other thermocouple on the other side of the chip). It's also possible to share one LT1025 among more than two thermocouples, but the farther the cold-junction compensation chip is located from the actual thermocouple termination, the less effective is its ability to compensate properly.

Signal Isolation

The level of signal isolation required depends upon the environment in which the sensor will be deployed. In many situations, the isolation requirements are relatively minimal since the sensor is not exposed to either a harsh or a particularly sensitive environment. In others, such as when thermocouples are placed near electrically driven heating elements or are used to measure the temperatures of humans, the application mandates a much higher level of isolation to protect either the sensing circuitry or the system being measured.

As one might expect, there are a variety of ways to isolate the thermocouple signals, with varying levels of cost, complexity, and signal degradation. The least expensive technique is to simply employ very high impedances between the powered sensing circuitry and the item being monitored. The low end from a cost perspective for this approach would be a simple op-amp buffer with high resistive inputs, which might offer an input impedance of several $M\Omega$ up to hundreds of $M\Omega$. Taking it a step further in both price and performance, an instrumentation amplifier such as the TI (formerly Burr-Brown) INA118 has an input impedance of 10^{10} Ω, and it can also withstand up to 40V of common-mode voltage on its inputs, which is extremely useful if the sensor has to operate in an environment in which there is the possibility that a common-mode voltage may be coupled onto the thermocouple signal.[17] As we've discussed before, this can occur if the thermocouple gets shorted to a power line (or other nonzero voltage); if the input circuitry can't handle the extra voltage, the sensor will be destroyed. By incorporating the over-voltage protection on the chip itself, the designer can reduce board space, assembly cost, and parts cost. Note, however, that 40V of common-mode protection is insufficient protection for most AC power signals, which may be 110 VAC–480 VAC in an industrial setting. If the system must protect against those voltage levels, additional protection circuitry is needed to knock the maximum input voltages down to levels that the chip can handle.

Finally, at the upper end of the price and performance spectrum are techniques that optically isolate the thermocouple signal from the rest of the system. This can be handled using either optically isolated analog amplifiers (very expensive) or by digitizing the nonisolated amplified thermocouple signal using an external ADC and then employing an optically isolated digital interface to transfer the digitized data to the processor (less expensive). Although optically isolating just the digital ADC interface is less costly, it may not be acceptable in situations in which the sensor must limit the currents and/or voltages to which the monitored system is exposed (for instance, in medical applications).

In this application, we've used the instrumentation amplifier isolation approach since it is easily constructed, relatively inexpensive, and provides excellent performance. Figure 6.8 shows the schematic for one channel of signal isolation using an INA326 instrumentation amplifier.

6.5   Firmware Implementation

To simplify the firmware development, the application makes extensive use of the Microchip 16-bit Language Tool libraries to access the dsPIC DSC's peripherals and to perform DSP functions such as filtering. The following sections describe individual elements of the firmware in greater detail to give the user a better understanding of what's involved in using the libraries and how their use affects the coding structure.

Signal Sampling

In this application, we are using the dsPIC DSC's internal 12-bit ADC to sample the thermocouple data. As you'll recall from Chapter 3, to configure the ADC module we need to know the following information:

1. the I/O port pins to use as analog inputs

2. the sampling rate

3. whether to use interleaved sampling or to generate a single interrupt after converting all of the signals during a given sampling period

4. whether we need to multiplex the results buffer

5. the format of the converted data

Since we're using the dsPICDEM, some of our hardware decisions have been made for us. The dsPICDEM board uses a 5V on-board regulator for V_{DD} and AV_{DD}, and it employs a 7.3728-MHz crystal operating in the XT with 4x PLL to clock the dsPIC device (essentially 29.4912 MHz). In addition, several of the analog inputs are dedicated to functions on the dsPICDEM board, namely:

Signal	*Dedicated*	*Description*
AN0/RB0	Yes	Programming Data (PGD) signal from ICD2
AN1/RB1	Yes	Programming Clock (PGC) signal from ICD2
AN2/RB2	No	Available for user
AN3/RB3	Yes	Digital potentiometer that has been low-pass filtered
AN4/RB4	Yes	Analog potentiometer RP2 – 0 to AV_{DD}
AN5/RB5	Yes	Analog potentiometer RP3 – 0 to AV_{DD}
AN6/RB6	Yes	Analog potentiometer RP1 – 0 to AV_{DD}
AN7/RB7	No	Available for user
AN8/RB8	Yes	Temperature sensor that has been low-pass filtered
AN9/RB9	No	Available for user
AN10/RB10	No	Available for user
AN11/RB11	No	Available for user
AN12/RB12	No	Available for user

Signal	*Dedicated*	*Description*
AN13/RB13	No	Available for user
AN14/RB14	No	Available for user
AN15/RB15	No	Available for user

Table 6.2. Analog Input Signal Assignments

Note that AN0 and AN1 are dedicated to PGD and PGC, respectively, only if the corresponding jumpers in jumper block J8 are installed. If those jumpers are left out, the application can use AN0/RB0 and AN1/RB1 as either analog inputs or digital I/O pins.

This application will use AN11 for the cold-junction compensation input and AN7 for the thermocouple input. The inputs are all single-ended signals referenced to ground (the instrumentation amplifiers having already taken care of the conversion from differential to single-ended signals), and they are scaled to use the full range of the analog supply voltage (AV_{SS} to AV_{DD}).

Digital Filter Implementation

Thanks to Microchip's DSP library and its dsPIC Filter Design™ software, the simple digital filtering that we'll be doing in this application basically becomes a matter of generating the correct filter coefficients in Filter Design and then implementing the FIR digital filter structure using the functions included in the DSP library. First, though, we need to look at the basic structure the library uses to process data, and we need to understand the limitations that using the library impose upon our application. These limitations include restrictions that apply to the library in general and those that are specific to its filtering routines.

The DSP library uses *fractional vectors* to store the data upon which it operates. Although it may sound impressive, a fractional vector is simply an array of two-byte data elements that use the 1.15 data format to represent each element's numeric value. The library has two requirements for a fractional vector: the array elements must be contiguous (i.e., linked lists are not permitted), and because they are 16-bit data values, the array must start on an even memory address. These minor restrictions are a direct consequence of the underlying dsPIC hardware and allow the library to employ the processor's specialized DSP engine to optimize mathematical operations. It's also important to note that, in general, the library functions are designed to operate on fractional vectors that have been allocated in the default RAM memory spaces (X-Data or Y-Data memory).[18] The filtering routines further restrict where certain data can be placed in memory; input and output data samples may reside in the default RAM memory spaces (X-Data or Y-Data), but filter coefficients can be in only X-Data or program memory, and the filter delay values *must* reside in Y-Data.[19]

Of course, all of the normal mathematical rules for handling vectors apply, as do common-sense programming policies such as having destination vector buffers that are large enough to handle the results of the specific mathematical operation. The reader should note that, for purposes of speed, the DSP libraries perform no checking of inputs or outputs for validity, and while the libraries do set the saturation and overflow flags, their functions do not make use of them. Unless the application's data specifically precludes this from happening, the programmer should check these flags following the conclusion of each call to the library.

For this application, we'll use the following parameters:

Sampling Frequency:	500 Hz
Passband Frequency:	15 Hz
Stopband Frequency:	25 Hz
Passband Ripple:	0.1 dB
Stopband Ripple:	3 dB
Filter Type:	Gaussian

Entering these parameters into the dsPIC Filter Design program and generating code for the resulting filter (stored in `SensorFilter.s`) provides us with the filtering data structure we need.

Data Analysis Implementation

The data analysis for this application is extremely straightforward: convert the filtered data to the corresponding temperature and then compare that temperature against an upper and a lower limit. While the processing is basic, these operations form the elements of not just a temperature monitoring system but also of any temperature control system. Never underestimate the power of simplicity…

Error-handling Implementation

As with the data analysis, we'll keep the error handling pretty simple in this application. Note that when we talk about error handling in this context, we mean unexpected conditions that adversely affect system operation, not the temperature out-of-range condition checked by the data analysis routines. The difference is subtle, but important. While an out-of-limits temperature condition may not be desired, it's perfectly valid for the system, and its occurrence doesn't prevent the system from operating properly. The error conditions that we discuss here are fundamentally different, however, because if they're not detected, the system will report invalid information and act upon it, possibly with disastrous consequences.

We'll check for two basic error conditions that commonly occur in temperature measurement systems: an open (broken) thermocouple and a reversed thermocouple. Each of the two conditions manifests itself slightly differently, and each affects the system operation to varying degree. For instance, an open thermocouple is basically a catastrophic failure that the application can detect but can't correct. In contrast, the system can compensate for a reversed thermocouple. Although, or perhaps because, an open thermocouple is a catastrophic failure, it's a relatively easy condition to identify; we're able to identify an open thermocouple simply by checking the raw input signal level against a threshold. Because we've attached pull-up and pull-down resistors to the thermocouple inputs, a broken thermocouple allows the input signal level to rise to near full-scale, which is clearly well outside the range of any valid thermocouple signal output.

Detecting a reversed thermocouple is more challenging because there are temperature ranges that produce a "valid" thermocouple output signal when the leads are reversed—i.e., temperatures whose corresponding outputs are within the voltage levels expected for the overall temperature range. The only way to know for certain whether the thermocouple leads are reversed is to look at the output voltage when the temperature changes. On a normal thermocouple signal that has been linearized, the output voltage increases (becomes more positive) as the temperature increases; if the thermocouple leads are reversed, the output voltage decreases (becomes more negative) with increasing temperature. If one has a thermal source available—for instance, when a heating element is paired with a thermocouple, the system can turn on the heater for a short period of time and check for a rise in the corresponding thermocouple's output voltage. If the voltage goes up when heat is added to the system, the wiring is correct (at least from a polarity perspective); if the voltage decreases, however, then the thermocouples are reversed. Obviously, if one had a cooling source (for example, in a refrigeration application), the same test could be applied, with the understanding that one would expect the voltage to *decrease* when the cooling element was activated.

The more difficult case occurs when no heat or cooling source is available, because then the only approach is to wait for the thermocouple output signal level to exceed the expected voltage limits in a negative direction (assuming we're applying heat to a system) or in a positive direction (in a cooling application). Once that condition has occurred, though, the firmware can flag the error and report it to the user.

Communication Protocol Implementation

The communication protocol used in all three sample applications was discussed in Chapter 4. We use UART 1 to transfer data via a standard RS-232 interface to the host system, and since the protocol's basic command structure supports everything we need for this particular application, we don't need to extend it at this point. To illustrate the way communications could be implemented using either a human-readable text protocol or a machine-readable binary protocol, the example code allows the designer to use either technique by simply uncommenting the appropriate line in the header file `ProtocolDef.h`.

The basic approach to processing data is to check for received data in the main processing loop using the function `CommIsRxPending()` and, if new serial data is available, to read it in with `CommGetRxData()` and then to process it by calling `ProtocolProcRxData()` with the newly received character. This function parses the data, and when a complete command message has been received, it calls the proper command handler through `ProtocolProcMsg()`. After identifying the command, `ProtocolProcMsg()` calls the corresponding low-level command-handler routine and transmits the appropriate response back to the host using the command-specific response routine. Once the response has been sent, `ProtocolProcRxData()` resets its parsing state machine to await the next message.

6.6   Summary

In this chapter, we've gotten our design feet wet with a brief dip in the sensor design pool. We're now able to digitize an analog sensor signal, filter it to remove unwanted noise, analyze the filtered data to extract some useful information, and communicate that information to other components in our system. As we develop more complex systems, we'll continue to use our framework as the foundation upon which we build. Let's now turn our attention to extending our framework a bit in the next chapter with our second design, a pressure/load sensor system.

Endnotes

1. *Temperature Technology Claimed to Eliminate Ambient Variations*, InTech magazine, December 1998.

2. Information on the Seebeck effect is widely available. One such resource is *Electromagnetic & Electrmechanical Machines, 2nd Edition*, by Leander W. Matsch. Copyright 1977 by Thomas Y. Crowell Company, Inc.

3. The reference information for both RTDs and thermistors is taken from http://www.omega.com/techref/measureguide.html and from http://rdfcorp.com/anotes/pa-r/pa-r_01.shtml.

4. *Emissivity* is usually denoted as a percentage between 0% and 100% or as a fraction between 0 and 1. In both cases, lower numbers represent less efficient emission of infrared radiation. The discussion of the relative emissivity values comes from http://www.omega.com/techref/measureguide.html.

5. In injection molding, molten plastic is usually referred to as *melt*, and the leading edge of the molten plastic is called the *melt front*.

6. For those of you who, like me, are not chefs, just picture a warm soggy mess.

7. The ITS-90 Tables of Thermoelectric Voltages and Coefficients can be downloaded from the NIST website at http://srdata.nist.gov/its90/download/download.html.

8. Data taken from the Omega Engineering, Inc. table found at http://www.omega.com/temperature/z/tcref.html.

9. *Clipping* occurs when the theoretical value of an analog output signal exceeds the actual output signal level because the circuit is incapable of delivery the desired output voltage. An example of this would be an amplifier circuit with power rails of ground and +5V and a nominal gain of 10. Any input signal between 0V and 0.5V will be amplified correctly, but input signals greater than 0.5V will have an output voltage of only 5V, since the circuit can only produce a maximum voltage equal to (or at least near) its power supply voltage.

10. A good indication that the ADC circuitry is excessively loading the sensor output is that the sensor output signal looks normal when viewed on an oscilloscope but then changes significantly when connected to the ADC input. Such changes may be in the form of reduced signal swing when the

sensor's input is varied or in the form of the sensor signal appearing to go to one of the two power voltages for the ADC.

11. *Murphy's Law* basically states that whatever can go wrong, will go wrong, and that given the choice between two items having a problem, the more critical of the two will be the one that fails.

12. *Temperature velocity* is simply the temperature change versus time and is analogous to the standard concept of velocity, which is simply the distance change versus time.

13. *Circuit Provides Cold-Junction Compensation*, by Mark Maddox and John Wynne. EDN Magazine, November 11, 2002, http://www.edn.com/article/CA260064.html.

14. The website http://www.daytronic.com/reference/aliasing.htm has a brief but informative discussion of the use of the Butterworth structure as an antialiasing filter.

15. The full version of the Digital Filter Design software is Microchip's part number SW300001. A reduced-feature version of the package also is available for significantly less, the key differences being that the less expensive version does not support as many FIR and IIR filter taps nor does it offer MATLAB support. In many cases, the "light" version of the software is completely adequate. The part number for the reduced-feature version is SW300001-LT.

16. Although the quotation "first, do no harm" is often mistakenly credited to Hippocrates, it is not found in the Hippocratic Oath. One possible source is from his work *Epidemics*, in which he states that physicians must "…have two special objects in view with regard to disease, namely, to do good or to do no harm" according to http://ancienthistory.about.com/od/greekmedicine/f/HippocraticOath.htm.

17. There are a great many instrumentation amplifiers from a variety of manufacturers that work well in this capacity. The TI part is simply one with which the author is well acquainted.

18. This note appears in the section, User Considerations, of the *16-Bit Language Tools Libraries* user's guide.

19. The requirements for memory space usage by the DSP filtering routines are found in the section, Fractional Filter Operations, of the *16-Bit Language Tools Libraries* user's guide.

Sensor Application— Pressure and Load Sensors

No pressure, no diamonds.
                    —Mary Case

With the temperature-sensing application we developed in the previous chapter, we gained a level of practical experience in the design of a sensor system. We now build upon that experience by extending the basic sensor framework to support load sensors and pressure sensors weight. It turns out that load sensors, which measure weight, and pressure sensors, which measure weight over a given area, have wide application in a tremendous number of diverse fields. Whether it's to count the number of tiny electronic parts by weight, to measure a patient's blood pressure, or to control the pressure exerted on a diamond-tip drill, load and pressure sensors are an integral part of our everyday life, whether we realize it or not.

7.1 Types of Load and Pressure Sensors

Before getting into the subject too deeply, it's important to realize that load sensors and pressure sensors are essentially identical. They both measure the load (or weight) on the sensor; the only difference is that a pressure sensor's output is scaled to divide that weight by the area of the sensor. For instance, one commonly available type of pressure sensor comes in an armored cylindrical package for use inside high-pressure environments that would otherwise destroy the sensing element. When the pressure sensor is monitored, its output signal reflects the force exerted upon the flat portion of the cylinder (assuming that it was installed properly). Knowing the diameter of the sensor, the application can then scale the measurement to compute the corresponding pressure.

An example may help. Assume that we're employing a cylindrical sensor like the one mentioned above, and further assume that the cylinder has a diameter of 0.125 inches (1/8 inch). If the sensor measures a force of 100 lbs on it, the corresponding pressure is given by the equation:

$$\text{Pressure} = \text{Load} / \text{Area}$$
$$= \text{Load} / (\pi \, r^2)$$
$$= 100 \text{ lbs.} / ((3.14)(0.125 \text{ inch})^2)$$
$$= 2{,}038 \text{ psi}[1]$$

If that same 100-lb force were exerted on a sensor with only half the diameter of the first sensor (i.e., a diameter of 0.0625 or 1/16 inch), the corresponding pressure would be

$$\text{Pressure} = 100 \text{ lbs} / ((3.14)(0.0625 \text{ inch})^2)$$
$$= 8{,}153 \text{ psi}$$

There are two types of sensors that are most commonly used to measure load and/or pressure in monitoring and control applications today: the strain gage and the piezoelectric sensor. While the discussion that follows is not an exhaustive treatise on the two sensor types, it should give us the information we require in order to interface to them and to get a feel for their application.

Strain Gages

In a nutshell, a strain gage is a device that changes its resistance in response to the force exerted upon it. This effect was first reported in 1856 by Lord Kelvin, but it wasn't until 1938 that it was put to practical use, since the change in resistance tends to be very small even for large changes in load. Interestingly, it was a circuit perfected way back in 1843 by the English physicist Sir Charles Wheatstone that allows us to measure those small changes accurately (see Figure 7.1).

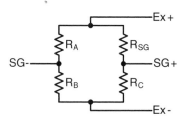

Figure 7.1. Schematic of a Wheatstone Bridge Strain Gage

In the circuit, Ex+ and Ex– are connections to an external excitation voltage that powers the circuit, and SG+ and SG– are the outputs whose differential voltage we measure. This arrangement of resistors produces an output voltage that is ratiometrically related to the values of the resistors in each leg of the bridge. If we match the resistances in the SG– leg (R1 and R3) to the corresponding resistances in the

SG+ leg (R_{SG} and R4), we'll get a zero output voltage. That may not seem to be a particularly startling fact, but it allows us to easily track changes in the resistance of a particular element because those changes show up in the form of a nonzero output voltage. As the reader has probably already anticipated, by placing the strain gage element as one of the resistors, we can measure accurately the changes in resistance caused by a force applied to the strain gage.

Load cells employ strain gages arranged in a bridge configuration to measure the force exerted upon a particular mechanical axis. The thermocouples we examined in the previous chapter are known as *passive sensing elements* because the change in the sensed parameter (in this case temperature) produces the sensor's output signal energy. In contrast, load cells are *active sensing elements* in which the change in the sensed parameter modulates an externally supplied *excitation* voltage. This gives the designer an additional degree of freedom since he can select excitation voltage characteristics that optimize parameter measurement, but it also requires additional circuitry to supply the excitation voltage and adds heat to the system that can degrade the sensor readings.

Like thermocouples, strain-gage load cells produce a small output voltage, usually expressed in terms of millivolts per volt of excitation voltage. This requires careful signal conditioning, but we can use the ratiometric characteristics of the bridge output to enhance our ability to measure these small signals. Because they are fairly easy to use and relatively inexpensive, strain-gage load cells are the most commonly used type, and it is that type of load cell that we will use in the application we develop in this chapter.

Piezoelectric Sensors

A second popular type of load sensor is the piezoelectric sensor, which is based upon the property of certain materials that produce a tiny electrical charge when a force is applied to them.[2] This output is extremely small, on the order of nanocoulombs,[3] and special circuitry known as a *charge amplifier* is required to convert these small changes in charge output to a voltage that can be measured by an external system.

Piezoelectric sensors are generally used in applications in which the load is dynamic—i.e., changing on a regular basis—because the response of the sensors to a rapidly changing load can be better than strain-gage load cells. That advantage may not necessarily translate into better readings, however, if the response of the system being measured is much less than that of the sensor itself. In addition, piezoelectric sensors are generally not as accurate in situations in which the load is relatively constant because the charge output tends to die out over time when presented with a constant force.

Unfortunately, charge amplifiers are expensive to purchase and difficult to build, which reduces the number of applications in which piezoelectric sensors can be used. However, in situations demanding the greatest frequency response, the additional cost for piezoelectric sensors may be mandated by the application requirements.

7.2 Key Aspects of Load Measurement

In general, the key aspects of load measurement are the same as those for any sensor system, namely range, resolution, and accuracy of measurement. While some of the challenges faced in load measurement as well as the signal characteristics are unique to this application, other considerations such as linearization and calibration are common sensor issues.

Range of Measurement

Because load-sensing applications cover such a broad range of load values, from micrograms to kilotons, the designer needs to determine the appropriate load range that the specific application will support. One key point that many engineers overlook (at least the first time) is the need to specify both the normal operating range and the worst-case dynamic range. A package-weighing system that is intended to handle a range of, say, 0 to 200 pounds may see an instantaneous force of several times that if a package is accidentally dropped from a height of a couple of feet. Not only is this a mechanical consideration (the system needs to be able to withstand the blow without being physically damaged), but it also requires that the signal-conditioning circuitry be such that the resultant instantaneous voltage signal does not damage any of the electronics.

As an example, we might design a pressure-sensing system with a normal operating range of 0–20,000 psi and a maximum range of 30,000 psi. This would be a reasonable range for measuring the cavity pressure (the pressure inside the mold) of an injection-molding machine. By choosing an operating range of 0–20,000 psi, we're basically saying that we want to maintain the greatest accuracy over that range, so we'll use those two points as our calibration points.

Resolution of Measurement

Assuming that we amplify the input signal so that the minimum and maximum parameter values correspond to AV_{SS} and AV_{DD}, respectively, the measurement resolution is obtained by dividing the maximum range by the number of ADC levels available to us (1,024 levels for a 10-bit ADC, 4,096 levels for a 12-bit ADC, or 65,536 levels for a 16-bit ADC). In this case, we're going to use the dsPIC30F6014A's on-chip 12-bit ADC, so we'll have 4,096 levels with which to work. Given the 30,000 psi range (remember, we have to use the maximum range we want to measure, not just the operating range), that translates to a measurement resolution of about 7.3 psi/level (30,000 psi/4,096 levels = 7.32 psi/level).

If we require additional accuracy, we would need to employ a higher resolution external ADC. For most injection-molding applications, however, this resolution is perfectly adequate.

Accuracy of Measurement

Another consideration is the accuracy of the measurement. This depends upon the entire signal chain from sensor through the ADC and requires careful attention to the analog signal-conditioning circuitry, the excitation voltage generator, the accuracy of the ADC, and any signal degradation due to thermal changes. Inaccuracy in the signal chain is cumulative, so each link in the chain is important. As a starting point, bridge-type load cells can be accurate to approximately 0.1%, though that depends upon the specific model selected.

Challenges

The challenges with bridge-type load cells are similar to those facing thermocouple signals, primarily the very low-voltage nature of the output signal (at least at low loads) and the degradation of the sensor's accuracy with changes in temperature.

Signal Characteristics

We've already touched on the load-cell signal characteristics a bit, namely that they are relatively low-voltage, on the order of a few millivolts to a few hundred millivolts. The signal range is affected by the excitation voltage level, but the designer usually has to balance the advantages of using a higher excitation voltage (which generates a greater output signal range) with the disadvantages of the associated self-heating effects.

While the frequency response of load-cell sensors are sensor-specific, in general they can be used to measure signals with frequency contents of tens of hertz up to hundreds of hertz. Using the conservative approach of a five-to-one ratio of sampling rate to frequency content, the designer should expect to sample at rates of 100 Hz to 1 kHz or more, depending upon the application. Obviously, applications in which the load is not changing during the measurement period can be sampled at far lower rates.

Thermal Compensation

The self-heating effect is most pronounced in applications that employ a large excitation voltage and that have limited thermal dissipation paths (such as in a hermetically sealed enclosure). Since we are using a low excitation voltage (5V) in our application, self-heating is not a concern; however, in situations that do have to worry about self-heating, the effect can be minimized by pulsing the excitation voltage using a low duty cycle. In such circumstances, the signal sampling must be synchronized with the duty cycle of the excitation voltage so that sampling only occurs when the excitation voltage is "on." This technique can also be employed to allow the use of a larger excitation voltage (say 10V), but care must be taken to ensure that the excitation voltage cannot damage any other circuitry. For instance, since this application uses the excitation voltage as the reference voltage, for the dsPIC's ADC, the excitation voltage is limited by the dsPIC DSC's electrical specifications for the reference voltage, even though the sensor output signal itself would be well within the valid input range for the ADC, even with a higher excitation voltage.

Linearization

Unlike thermocouple sensors, linearization is not a particularly significant issue for strain gage-based load cells, since the bridge's output is pretty linear over the load cell's operating range. This greatly simplifies the designer's task, since he can dispense with any special linearization hardware or software.

Calibration

Calibrating a strain gage-based load cell can be physically difficult depending upon the application. In some situations, particularly those for weighing, the designer can assume that the force exerted by the load will be perpendicular to the ground. Assuming that the maximum load is "reasonable" (an admittedly loose definition), the user may be able to calibrate the sensor system by simply using two standard weights near the upper and lower ends of the expected operating range. The system would log the measured value of the sensor output voltage at each calibration point and then compute a straight-line calibration gain and offset value as we did for the thermocouple-based system.

Unfortunately, we're often not in a position to calibrate the system so easily, either because the loads are not "reasonable" (e.g., it's not particular feasible in most circumstances to swap out a 100-pound load and a 50,000-pound load as calibration points) or because the physical characteristics of the system being measured preclude the use of calibrated physical loads. An example of the latter condition occurs in injection-molding machines, which often have vertically mounted sensors that measure the pressure of the plastic melt inside the mold. It's not possible (or at least feasible) to somehow apply a calibrated 20,000-psi physical load to such a sensor, so we have to find an alternative approach.

In these situations, a fairly common technique is to apply a known resistance across the load cell output and then to use the resulting output voltage as the upper calibration point (the lower calibration point being the unloaded sensor output signal). Although this may sound crude, it works fairly well in practice and often is the best option available, whether we like it or not.

Sources of Noise

Noise can get onto the load-cell output signal through a number of different sources. AC power noise can be coupled onto the excitation voltage leads going into the sensor, onto the output signal leads coming from the bridge, or both. Assuming that the leads are close to one another (and ideally constructed as twisted pairs), the coupled noise is common-mode in nature and thus can be removed for the most part by a differential amplifier for the signal input and by the ratiometric nature of the ADC measurement. That's important, because a 50 Hz or 60 Hz power signal can be right in the middle of the valid frequency content of the signal we're measuring.

If the inherent noise-rejection characteristics of the circuitry don't remove the AC noise sufficiently, the designer can always employ a digital narrow-band notch filter with the notch centered at either 50-Hz or 60-Hz (depending on the power frequency). Although this adds to the complexity of the software and degrades the measured signal somewhat, such a filtering arrangement may be necessary depending upon the application and the electrical environment.

Error Conditions

Load cells are subject to several possible error conditions, including:

1. a broken lead to/from the sensor,
2. a damaged bridge element in the sensor,
3. reversed excitation voltage leads, and
4. reversed output voltage leads.

All of these conditions can be detected by measuring the resistance across the various sensor leads, but allowing the system itself to do so requires some additional circuitry and reduces the number of ADC channels available for load-cell output signals, since some of those channels must be used for diagnostics. Of these conditions, the first three are "fatal" in the sense that the system cannot operate if they exist. It's possible to recover from the fourth condition (reversed output voltage leads) if it has been detected.

7.3 Application Design

The application design is a fairly simple weighing device that uses a load cell to accurately measure the weight of an item between 0 and 50 pounds. To simplify the calibration, we'll use the known load resistance approach, switching in a resistor of a known value to serve as the upper calibration point.

System Specification

The system uses the Microchip dsPICDEM v1.1 board as the hardware platform and performs the following tasks:

1. Measure a load of between 0 and 50 pounds, with a resolution of 0.25 ounce.

2. Maximum load of 64 pounds.

3. Sample a single load channel at a rate of 1 kHz (assumes a maximum frequency content of 200 Hz).

4. Filter the sampled data to remove 60-Hz power-line noise.

5. Allow the user to perform the following functions via the RS-232 serial port running at 38.4 Kbps, 8 data bits, 1 stop bit, no parity, and no flow control:

   a. perform calibration of the unit,

   b. specify an upper and a lower limit for the measured weight.

6. Report the measured load every second via the serial port. Load reports are to be in text.

7. Report out-of-limit alarm conditions by lighting LED1 on the demo board if the measured temperature is less than the lower limit and by lighting LED2 if the temperature exceeds the upper alarm limit.

As with the temperature-monitoring system, if the information is to be read by other electronic components in the system rather than by humans, a faster binary data protocol would be appropriate, and as with the thermocouple example, one is provided in the sample code.

Sensor Signal Conditioning

The block diagram for the load-cell interface is shown in Figure 7.2, and it consists primarily of an instrumentation amplifier with gain to condition the load cell's output signal, the excitation voltage source, which in this case is the system's 5V power, and a pair of cascaded second-order Butterworth antialiasing filter to limit the frequency content of the amplified signal prior to sampling by the ADC. In addition, the interface includes a relay that allows the processor to switch in a resistor of a known value as the calibration shunt.

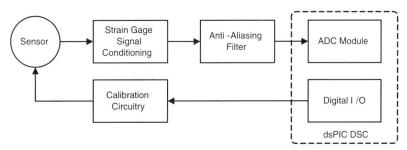

Figure 7.2. Block Diagram of Load-cell Interface (Single Channel)

Note that we're able to use the same differential amplification circuitry (albeit with a different gain) and the same Butterworth filter topology (though with different component values to accommodate the broader bandwidth) that we used to condition the thermocouple signals. The schematic for the actual circuitry is shown in Figure 7.3. As with the temperature measurement example, the schematic is for a 3.3V system; when running on the dsPICDEM 1.1. GPDB, change all references for 3.3V to 5V, all those for -3.3V to -5V, and that for 1.65V to 2.5V.

Digital Filter Analysis

Because the AC power noise is considered to be in-band with the signal we're measuring, we need to use a sharp notch filter to remove it. The frequency response of the filter we'll use is shown in Figure 7.4.

Data Analysis Algorithms

The data analysis algorithms are extremely simple and basically consist of the same analysis algorithms employed by the temperature monitor with minor changes. The algorithms simply check each filtered sample to determine whether the sample value is outside either the upper or the lower alarm limits and then lights the appropriate LED if that is indeed the case.

Load Cell Interface Circuitry

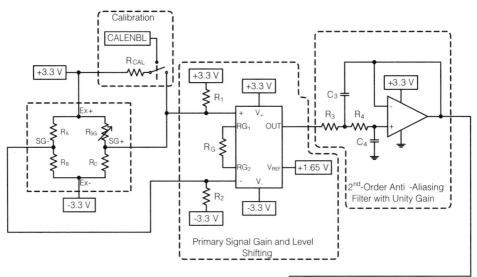

NOTES:

1. R_1 and R_2 provide an input common-mode current path and a way to determine whether the strain gage is actually connected to the circuit. Resistor values should be identical and be approximately 1 MΩ to avoid loading the input signal with too low an input impedance (particularly during calibration).

2. Gain through the instrumentation amplifier is controlled by R_G according to the equation
$$\text{Gain} = 1 + (50 \text{ K}\Omega \ / R_G)$$

3. Anti-aliasing filters have a Butterworth filter frequency response. Gain is included in the final filter stage to compensate for the instrumentation amplifier's inability to produce output voltages that are near either power supply rail. The gain in the final stage is given by the equation
$$\text{Gain} = 1 + (R_8 \ / R_7)$$

Figure 7.3. Schematic of Load-cell Interface

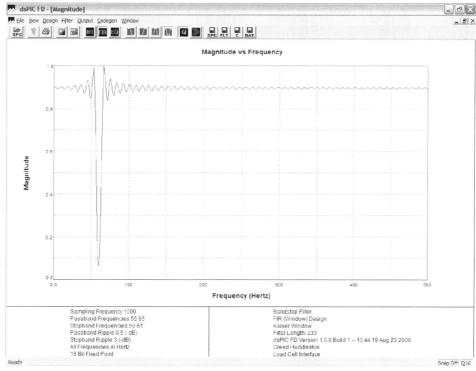

Figure 7.4. Response of Notch Filter for AC Power Noise Removal

7.4   Firmware Implementation

As with the temperature-monitoring system, the firmware that implements the signal sampling, filtering, analysis, and error handling for the weigh scale relies on Microchip's 16-bit Peripheral Library and its DSP Library to encapsulate many of the hardware interface functionality and to provide a consistent, tested set of digital signal processing routines for filtering.

Signal Sampling

The load-cell signal on analog input pin AN7 is sampled every millisecond using the autosample mode and the dsPIC's Timer 3 to trigger the conversions. To maximize the resolution for this application, we'll assume that the input signal is unipolar (i.e., we'll require that the user ensure that the excitation voltage can't get reversed) with an input range of 0 to 5V, so we can use the unsigned 1Q15 format for the data from the ADC. Also, to reduce the overhead associated with a fairly quick sampling rate, we'll interrupt only after every eight samples have been added to the buffer. This number can be adjusted down to increase the system response or up to further reduce the interrupt-associated processor overhead.

The code that configures the ADC sampling parameters is found in the function ADCInit() in the module ADCIF.c. It uses the OpenADC12() and the OpenTimer3() routines from the 16-bit Peripheral Library to set up both the ADC and Timer 3 as shown in the following code fragments from ADCIF.c and Timer.c.

Code Example 7.1. Signal Sampling Configuration Code

From ADCIF.c:

```
Uint8
    ADCInit(void)
    {
    // Local Variables

    Uint16
        ui16ADCON1,        // Configuration data for ADCON1 register
        ui16ADCON2,        // Configuration data for ADCON2 register
        ui16ADCON3,        // Configuration data for ADCON3 register
        ui16ADPinConfig,   // Configuration data for ADPCFG (pin
                           //    configuration)
        ui16ScanSelect;    // Configuration data for ADCSSL (channel
                           //    scan selection)

    //  Turn off the ADC module and disable the ADC
    //  interrupt to ensure that the configuration
    //  completes without interruption

    CloseADC12();

    //  Setup the ADCON1 register configuration
    //  data, of which the most important is the
    //  data format, the clock source, and the
    //  Auto Sampling mode

    ui16ADCON1 = ADC_MODULE_ON      & // Enable the ADC module
                 ADC_IDLE_STOP      & // Stop conversions during IDLE state
                 ADC_FORMAT_FRACT   & // Generate unsigned fractional data
                 ADC_CLK_TMR        & // Use Timer 3 to trigger conversions
                 ADC_AUTO_SAMPLING_ON; // Enable the Auto Sampling mode
```

```
//  Setup the ADCON2 register configuration
//  data, which includes the ADC reference
//  voltages, the number of samples between
//  interrupts, and the buffer data order
//  specification.

ui16ADCON2 = ADC_VREF_AVDD_AVSS    & // Use AVdd and AVss as ADC reference
             ADC_SCAN_ON           & // Enable channel scan mode
             ADC_SAMPLES_PER_INT_8 & // Gather 8 samples before interrupting
             ADC_ALT_BUF_OFF       & // Don't alternate the sample buffer
             ADC_ALT_INPUT_OFF;    & // Don't interleave the input


//  Setup the ADCON3 register configuration
//  data, which includes

ui16ADCON3 = ADC_SAMPLE_TIME_31  &  // Sample for 31 Tad
             ADC_CONV_CLK_SYSTEM &  // Use system clock for conversions
             ADC_CONV_CLK_32Tcy;    // Allow 32 Tcy for conversion


//  Setup the ADCSSL register configuration
//  data, which specifies which channels are to
//  be scanned by the ADC. Note that the way
//  the library is implemented, we specify the
//  ANx signals that we do NOT want to scan in
//  this list. In this case, we leave out AN7
//  since that's the only one we want to sample.

ui16ScanSelect = SKIP_SCAN_AN0  & SKIP_SCAN_AN1  & SKIP_SCAN_AN2  &
                 SKIP_SCAN_AN3  & SKIP_SCAN_AN4  & SKIP_SCAN_AN5  &
                 SKIP_SCAN_AN6  & SKIP_SCAN_AN8  & SKIP_SCAN_AN9  &
                 SKIP_SCAN_AN10 & SKIP_SCAN_AN11 & SKIP_SCAN_AN12 &
                 SKIP_SCAN_AN13 & SKIP_SCAN_AN14 & SKIP_SCAN_AN15;


//  Make sure that the pins associated with
//  the analog channel(s) we're using have
//  been configured as analog inputs. This
//  application uses AN7 as the signal input,
//  and the dsPICDEM board dedicates AN0 and
//  AN1 to the ICD2 interface, AN3 to the
//  digital potentiometer input, AN4-AN6 to
//  the analog potentiometer inputs, and AN8
//  to the temperature sensor input.
```

```
ui16ADPinConfig = ENABLE_AN3_ANA & ENABLE_AN4_ANA & ENABLE_AN5_ANA &
                  ENABLE_AN6_ANA & ENABLE_AN8_ANA;


// Actually configure the ADC module. Note
// that this will enable the ADC module, but
// the associated interrupt must be enabled
// by a call to EnableIntADC() after the rest
// of the system has been initialized

OpenADC12(ui16ADCON1, ui16ADCON2, ui16ADCON3, ui16ADPinConfig,
        ui16ScanSelect);


// Initialize Timer 3 to generate a 1 msec
// interrupt to trigger the ADC sampling

Timer3Init(ADC_SAMPLE_RATE);


return ST_OK;
}
```

From `Timer.c`:

```
void
  Timer3Init(Uint16 ui16SampleRate)
  {
// Local Variables

Uint16
    ui16Period,                 // Timer period in counts
    ui16TimerCfg;               // Timer module configuration data

// First, turn off Timer 3 and its associated
// interrupt so we can complete the initialization
// without being interrupted

CloseTimer3();

// Compute the timer period in counts based
// on the instruction clock frequency. The
// instruction clock frequency is computed
// by multiplying the crystal frequency FOSC
// by the phase-locked loop scaler PLL and
// then dividing by 4 since each instruction
// takes four system clock cycles.
```

```
ui16Period = (Uint16)(((FOSC * PLL / 4) / ui16SampleRate) + 1);


//  Configure the timer itself


ui16TimerCfg = T3_ON            &    // Enable Timer 3
               T3_IDLE_STOP     &    // Turn off Timer 3 in IDLE state
               T3_GATE_OFF      &    // Not gating the timer
               T3_PS_1_1        &    // Set the prescaler to 1:1
               T3_SOURCE_INT;        // Use internal clock for timer


OpenTimer3(ui16TimerCfg, ui16Period);    // Configure the timer


return;
}
```

The interrupt handler is the routine ADCISR(), also found in the ADCIF.C module. The handler clears the ADC interrupt flag. Failure to clear the ADC interrupt flag will cause the processor to vector back into the handler as soon as the function returns, essentially dooming the application to a death spiral of constantly servicing the interrupt.

Code Example 7.2. ADC Interrupt-handler Code

From ADCIF.c:

```
void __attribute__((__interrupt__))
    ADCISR(void)
    {
//  Local Variables

fractional volatile
    *pfrADCBuff,            // Pointer to ADC data buffer
    *pfrInputSignal;        // Pointer to global input signal data
                            //   buffer


Uint8
    ui8DataCount;           // ADC data index

//  Clear the ADC interrupt flag and

IFS0bits.ADIF = 0;          // Clear the ADC interrupt flag

//  Copy the A/D conversion results to the buffer
```

```
//  g_frSensorSignal[]. Note that the samples
//  from each channel are contiguous, so they will
//  have to be placed into separate signal arrays
//  the samples are to be filtered. We don't do
//  that here in order to minimize the time spent
//  in the ISR.

pfrADCBuff      = &ADCBUF0;       // Point to the first entry in the ADC
                                 //   data buffer
pfrInputSignal = g_frInputSignal; // Point to the first entry in the input
                                 //   signal data buffer

for (ui8DataCount = 0; ui8DataCount < ADC_SAMPLE_COUNT; ui8DataCount++)
    *pfrInputSignal++ = *pfrADCBuff++;   // Copy the next ADC sample to the
                                 //   global input signal buffer

//  Set the Filter Event to signal the main
//  processing loop that we have new data to
//  filter

g_vui16SysEvent |= EVT_FILTER;
}
```

Digital Filter Implementation

The digital filter implementation is nearly identical to that for the temperature sensor discussed in Chapter 6, the only difference being the filter coefficients. As in that case, the filter used here is an FIR filter that requires the application to create and initialize an FIRStruct data structure to maintain the filter's associated state variables. Filtering of new data is simply a matter of calling the DSP library routine FIR() with the new data samples.

For this application, we'll use a bandstop filter with the following parameters:

Sampling Frequency:	1000 Hz
Passband Frequencies:	55, 65 Hz
Stopband Frequencies:	59, 61 Hz
Passband Ripple:	0.5 dB
Stopband Ripple:	3 dB
Filter Type:	Kaiser

Entering these parameters into the dsPIC Filter Design program and generating code for the resulting filter (stored in `SensorFilter.s`) provides us with the filtering data structure we need. Note that the filter length is much longer than that for the thermocouple application (233 taps vs. 51 taps) because the filtering requirements are much more stringent.

Data Analysis Implementation

The data analysis algorithm is handled by the routine `AnalyzeData()` in the `Analysis.c` module. This admittedly simple function merely examines the filtered data on a sample-by-sample basis and turns the lower and upper limit LED's on or off whenever the samples are beyond the specified limits.

Code Example 7.3. Data Analysis Code

From `Analysis.c`:

```
Uint16
    AnalyzeData(void)
    {
    //  Local Variables

    float
        fSensorValue;                       // Current scaled sensor value

    PSensorCfg
        psenSensorCfg;                      // Pointer to sensor configuration

    Uint16
        ui16SensorIndex;                    // 0-based sensor index

    //  Check the sensors one at a time to see
    //  whether the sensors' current values
    //  exceed any enabled alarm limits

    psenSensorCfg = g_senSensorCfg;         // Point to first sensor configuration

    for (ui16SensorIndex = 0; ui16SensorIndex < MAX_CAL_SENSORS;
        ui16SensorIndex++, psenSensorCfg++)
        {
        //  Convert the current filtered sensor
        //  reading into its corresponding parameter
```

```
// data value

fSensorValue = ScaleData(g_frFilteredSensor[ui16SensorIndex],
                         psenSensorCfg->fGain,
                         psenSensorCfg->fOffset + 0.0005);
g_fSensorValue[ui16SensorIndex] = fSensorValue;

// Are we checking the lower alarm limit
// and, if so, is the sensor's filtered
// data value less than the lower limit?

if ((psenSensorCfg->ui16Flags & SENFLG_ENBL_LOW_ALM) &&
    (fSensorValue < psenSensorCfg->fAlarmLevel[ALARM_LOWER]))
    psenSensorCfg->ui16Flags |= SENFLG_LOW_ALM_ST;
                                    // Alarm limit exceeded
                                    //   so set its flag

// Are we checking the upper alarm limit
// and, if so, is the sensor's filtered
// data value greater than the upper limit?

if ((psenSensorCfg->ui16Flags & SENFLG_ENBL_HI_ALM) &&
    (fSensorValue > psenSensorCfg->fAlarmLevel[ALARM_UPPER]))
    psenSensorCfg->ui16Flags |= SENFLG_HI_ALM_ST;
                                    // Alarm limit exceeded
                                    //   so set its flag
}

return ST_OK;
}
```

Error-handling Implementation

Error handling for the application can be relatively simple or it can be more complex, particularly if the application calls for testing the resistance of each leg of the bridge. Doing so requires pinning out the load-cell bridge connections to individual analog input lines, significantly increasing the complexity of the circuitry and diminishing the number of load-cell channels that can be monitored by a single dsPIC DSC using the processor's internal resources.

The error-handling code is located in the routine CheckForErrors() in Analysis.c, and it consists of checking for a shorted or an open bridge connection, conditions that are indicated by an extremely low or a nearly full-scale value, respectively. If either condition is detected, the code sets a flag that blinks the alarm LED (LED 2). Once the condition has been cleared, the application turns the alarm LED off.

7.5 Summary

Now that we've completed two applications using the dsPIC's ADC and digital filtering, we'll tackle a third application that acquires data using an entirely different approach: counting the number of transitions on a timer/counter external clock input to determine fluid flow through a turbine.

Endnotes

1. psi = pounds per square inch.

2. This property is known as the piezoelectric effect.

3. 1 nanocoulomb = 10^{-9} coulombs.

Sensor Application—Flow Sensors

A process cannot be understood by stopping it. Understanding must move with the flow of the process, must join it, and must flow with it.

—Frank Herbert, Dune (First Law of Mentat)

Rounding out our discussion of interfaces to popular sensors is an examination of flow sensors, those devices that allow us to measure the flow of a substance, usually a liquid, through a system. Although there are a number of ways to define flow, including *mass flow*, *volume flow*, *laminar flow*, and *turbulent flow*, the key point is that we want to measure the amount of the substance that is flowing to accomplish a particular purpose.[1] This usually means that system designers are interested in *mass flow* (the mass per unit time that passes a particular point in the system), but if the substance is a liquid whose density is essentially constant (at least through the point or points of interest in the system), the designer can substitute a measurement of *volume flow* (the volume per unit time) for mass flow with little effect on system performance. This occurs in many situations and is the subject of this final design example.

Where might we find applications of flow measurement? For most people, the two most prolific and prosaic examples are in the form of their electricity and water bills, which are based on the flow of those two substances into their home or office. Huge systems such as oil pipelines and small systems for monitoring the breathing of infants in neonatal intensive-care units employ flow sensors to monitor and control critical operational parameters. It is no exaggeration to state that commerce itself relies on reliable flow-measurement systems to ensure the accurate delivery of thousands of types of fuels, foods, and other substances worldwide.

8.1 Types of Flow Sensors

Although we'll describe two common types of flow sensors, the turbine sensor (which measures volume flow) and the *gravimetric sensor* (which measures mass flow), our application will focus on the turbine sensor for two reasons. The first is that in-line turbine sensors are widely used, so an understanding of them is extremely helpful to the designer. The second, and perhaps the more important, reason is that turbine sensors offer a type of sensor output that we have not used in the previous two applications, namely a *frequency-based* output signal whose frequency increases with increasing flow. As we'll see, this new output signal type has both advantages and disadvantages, and knowing how to deal with frequency-based sensors is a valuable tool to have in the designer's kit.

Turbine Sensors

Typically used with fluids or gases, the turbine sensor looks like a little fan or propeller, with the axis of the fan aligned with the direction of flow and the blades of the fan covering the entire cross-section of the flow path as shown in Figure 8.1. Not only does this blade arrangement allow the flow to exert as much pressure as possible on the sensor (very helpful in low-flow conditions), it also ensures that the flowing substance sees a constant pressure as it passes the sensor, which guards against possible material breakdown under high-flow conditions.

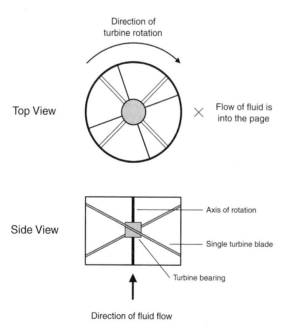

Figure 8.1. Example of a Turbine Flow Sensor

Because the fan axis is in-line with the material flow, it can be difficult to measure the axial rotation directly, so turbine sensors often have windows that allow the blades to be observed from the side. By counting the number of times a blade passes a fixed point on the window in a given period of time and knowing both the density of the material and the cross-sectional area of the flow path, the sensor can determine the volume of fluid that has passed through the sensor.

At first, monitoring the blade rotation might seem to be a daunting task, given the electrical isolation of the blade from the rest of the system. Fortunately, in many cases we can simply illuminate the window with a source that emits light at a frequency which the blade (but not the flowing material nor the sensor window) reflects well and then measure the reflected signal to get an output that rises and falls as the blade passes the point of illumination. A sample of such a reflectance signal is shown in Figure 8.2, and by counting the peaks of the signal in a given time period, we can determine how fast the sensor blades are rotating. We'll examine the issues we face with this arrangement in the section, Challenges of Flow Measurement.

Example Turbine Flow Sensor Output

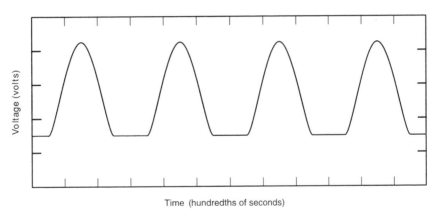

Time (hundredths of seconds)

Figure 8.2. Example Turbine-flow Sensor Output Signal

Unlike temperature sensors and load-cell sensors, which are primarily electrical in nature, good turbine-flow sensor design requires a high degree of mechanical as well as electrical skills to obtain optimum results. While we'll focus on the electrical aspects of the sensor interface here, the reader should be aware that the mechanical aspects of flow are important as well, and instrumentation of a system should include an understanding of the effect that the insertion of the sensor(s) will have on the system being monitored.

Gravimetric Sensors

Unlike turbine-flow sensors, which measure the volume of material that passes a cross-section of area during a given time, gravimetric sensors measure the actual mass of material that passes a cross-section during a given time. Although sometimes more accurate than a corresponding volume flow measurement, gravimetric flow sensors tend to be significantly more expensive than their turbine flow brethren (by as much as an order of magnitude), so their use tends to be limited to applications that require an extremely high level of accuracy and that can afford the added cost, such as in a calibration laboratory.

8.2 Key Aspects of Flow Measurement

Many of the key aspects of flow measurement are shared by all sensor systems, such as the range and resolution of measurement, while others are unique to flow sensors themselves, such as the challenges faced in low-flow and high-flow environments.

Range of Measurement

The range of measurement is dependent upon the physical characteristics of the material being monitored, the properties of the turbine being used to perform the measurement, and the nature of the system in which the material is flowing. Above a certain flow, the material under observation can be damaged as the shear forces literally begin to tear the material apart. In addition, the nature of the material can seriously affect the life of the sensor itself, particularly the blades, since abrasive materials under high flow can quickly wear away the blades. This adversely affects the quality of the sensor measurements, because the reduced blade area allows additional material to flow past the sensor without being measured and also decreases the ability of the sensor to operate in low-flow conditions due to the reduced surface area of the blades (which reduces the rotational pressure on the sensor).

The characteristics of the turbine determine both the lower and the upper limits of flow that can be measured, since too low a flow will prevent the turbine from moving (the pressure of the flow being insufficient to overcome the friction of the blades against its axis of rotation), while too high a flow can physically deform the blades, generating measurement errors and creating wear in the blade-axis junction, which produces blade "wobble" that degrades the resultant measurements. This latter problem is so pervasive that we'll deal with it in the section, Challenges of Flow Measurement, and specifically in the subsection, Sources of Noise.

Finally, the system itself will determine the range requirements for the flow measurements. In situations that require measurements of flow levels that exceed the capabilities of the sensor being used, a "sampling" tube of known size that diverts a portion of the flow for measurement purposes can be employed, with the measured flow through the sampling tube correlated to a corresponding flow through the main system.

All that being said, some typical turbine-flow sensor families support measurement ranges of 0.2 gallons per minute (GPM) up to 60 GPM.[2] Specialized sensors can extend this range to much larger flows.

Resolution of Measurement

Unlike the previous two sensor systems we've studied, which both generate a voltage signal that is directly correlated to the value level of the parameter of interest, the turbine flow sensor generates a signal whose *frequency*, not its voltage, correlates to the parameter being measured. To acquire the signal, we'll use a slightly different analog front end to convert the analog reflectance signal to a digital frequency signal and use that signal to drive one of the dsPIC DSC's Timer/Counter modules. This module counts the number of transitions on its clock input signal and, combined with a second timer module that has a resolution of the system-clock period (33.9 ns for the dsPICDEM v1.1 configuration), we will be able to make measurements well within the requirements of any real-world turbine-flow sensor.

Accuracy of Measurement

Turbine-flow sensors are capable of an accuracy within 2% or better of the full-scale reading, and their measurements are highly repeatable, provided that the turbine's rotational path is clear and that the turbine itself is undamaged by wear or age. Usually, the accuracy of the measurements degrades near the extremes—i.e., under extremely low flow or very high flow conditions—and this degradation worsens rapidly if the sensor's turbine assembly develops significant wear, particularly in the bearing that mounts the turbine to the axis about which it rotates. Fortunately, there are techniques we can employ that can ameliorate this loss of accuracy to an extent, and while the methods can't actually prevent the underlying causes of the problem, they can at least allow the system's performance to degrade gracefully. Often this gives the user an opportunity to monitor the system and to schedule required maintenance in advance rather than having to react to the problem on short notice.

Challenges of Flow Measurement

Turbine-flow sensors face the most serious problems at the lower and upper limits of flow. Low-flow conditions present a challenge because, below a certain level, the pressure of the flowing material on the turbine blades is insufficient to overcome the friction of the blades on the axis about which they're rotating. In this situation, the material can ooze around the nonrotating blades, allowing flow even though the sensor reads no flow. This condition can be alleviated by adding more blades to the turbine, since the additional force experienced by the added blades further reduces the flow level that can be measured, but doing so also increases the frequency output of the sensor (since more blades means more blade transitions per rotation), which increases the overhead required to acquire the data.

At the other end of the spectrum is that, under high-flow conditions, the reflectance signal tends to flatten out, with the peaks and valleys decreasing significantly in amplitude. This requires that the comparator circuitry used to convert the analog reflectance signal to a digital value support dynamic adjustment of the comparison threshold, and the firmware must be able to monitor the extremes of the analog signal in order to set that threshold correctly.

Over time, the bearing that mounts the sensor blades about the axis of rotation will tend to wear, enlarging the mounting hole about the axis. The enlarged mounting hole allows the blades to wobble around the axis and to ride up and down on it, which means that the blades passing the illuminated reference point may tilt between successive rotations, causing the reflectance peaks to shift back and forth somewhat about the "true" period. Depending on the type of material used and the flow level, bearing wear can arise fairly quickly and is a serious problem when it occurs.

Signal Characteristics

The analog reflectance signal is relatively large, on the order of hundreds of millivolts to volts, so amplification is not really an issue, although loading of the sensor signal and the need to remove common-mode noise may recommend the use of an instrumentation amplifier to buffer the sensor signal before further processing.

The frequency content of the turbine signal is easily computed as the number of blades times the maximum rotational speed supported by the sensor, and the minimum sampling frequency is twice that to meet the Nyquist criterion. add "Note that in this application, we're sampling the signal only to identify it minimum and maximum signal levels and not to extract parameter information from the sampled voltage signal. The task of parameter value extraction is handled by the counter we'll use, and our main consideration for the counter is to ensure that it can't overflow during the data acquisition period.

Material Density Compensation

To convert the flow-volume measurement to a mass-volume measurement, the sensor has to apply a material density factor. This factor is simply a material-dependent value that is multiplied by the volume-flow measurement to obtain the corresponding mass flow, with the basic assumption being that the density of the material is constant over the range of flow being measured. If that is not the case, the material density compensation factor can be applied in a piecewise-linear fashion similar to that which was used with the thermocouple measurements.

Linearization

In most cases, no special linearization algorithm is required since the frequency usually increases linearly with flow. The linearity tends to be somewhat worse at the two extremes of low flow and high flow, and it may be necessary in some circumstances to perform a simple three-part piecewise linearization to maintain accuracy near the limits of the sensor.

Calibration

The calibration needs of a turbine-flow sensor are fairly minimal, since the output signal characteristics of the sensor are dictated by the sensor's mechanics rather than its electrical response. Generally, the calibration curve for a sensor is fixed for a given sensor model and does not require any calibration in the field.

Sources of Noise

There are three primary sources of signal degradation with a turbine-flow sensor: bearing wear, turbidity of the material, and electronic noise in the analog portion of the sensor interface. The cause and effects of bearing wear have already been discussed in detail, and the main problem with material turbidity is that it can significantly reduce the reflected signal from the turbine blades, which adversely affects the ability of the sensor system to discern valid blade transitions through the reference zone.

Electrical noise is also a problem, although generally not to the extent seen in the previous two applications. The primary area affected by electrical noise is the comparator input, where noise can generate false comparator output transitions that result in a measured flow value that's higher than the true flow.

We'll explore the techniques to address the problems caused by these noise sources in the section, Signal Conditioning, but within limits, all three of these conditions can be handled satisfactorily.

Error Conditions

About the only hardware-error condition that can be detected for a turbine-flow sensor is a broken sensor lead, and that only under certain circumstances. Additional system problems such as excessive flow can be detected, but those reflect an actual, valid system condition (although they may be beyond the capabilities of the sensor) and thus are treated under the data analysis section rather than here.

The best way to detect a broken sensor lead is to place a sense resistor from the sensor output to ground and then to periodically monitor the voltage across the sense resistors to verify that it meets a minimum level consistent with what is expected out of the sensor. Depending upon the output level of the turbine sensor itself, a high-impedance amplifier may be required between the sense resistor and the ADC input.

8.3   Application Design

As with the previous applications we've developed, the flow-sensor system will be fairly simple but will illustrate the issues important to flow measurement. Our system will measure flows of between 0.2 GPM (roughly 0.8 liters per minute or LPM) and 8 GPM using a turbine flow sensor. Measurements above or below a set of limits will light a corresponding Alarm Limit LED, and a broken or disconnected signal lead will blink a Hardware Error LED.

System Specification

The system uses the Microchip dsPICDEM v1.1 board as the hardware platform and performs the following tasks:

1. Measures a flow rate between 0.2 GPM and 8 GPM with an accuracy of 0.01 GPM.

2. Allows the user to perform the following functions via the RS-232 serial port running at 38.4 Kbps, 8 data bits, 1 stop bit, no parity, and no flow control:

   a. set the flow factor value for any sensor or material,

   b. specify an upper and a lower limit for the measured weight.

3. Report the measured flow every second via the serial port. Flow reports are to be in text.

4. Report out-of-limit alarm conditions by displaying a sensor value of "----" and lighting LED1 on the demo board if the measured flow is less than the lower limit and by displaying "++++" and lighting LED2 if the flow exceeds the upper alarm limit.

5. Report hardware-error conditions (broken sensor lead) by blinking LED3 on the demo board.

As with the previous two applications, if the information is to be read by other electronic components in the system rather than by humans, a faster binary data protocol would be appropriate, and one is included in the example software.

Sensor Signal Conditioning

Because of the nature of the sensor signal, the signal conditioning is somewhat different from that performed in the two previous applications. In particular, the sensor output signal is buffered and passed through a comparator to generate the corresponding binary digital signal that can be measured using the dsPIC DSC's external counter function. To ameliorate problems with electrical noise, the comparator can use hysteresis and an adjustable thresholding mechanism that allows the dsPIC DSC to compensate for low-level sensor signals at higher flow rates. A block diagram of the analog signal-conditioning circuitry is shown in Figure 8.3.

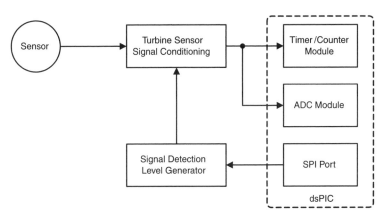

Figure 8.3. Block Diagram of Flow Sensor Signal Conditioning

Digital Filtering Analysis

The system employs a low-pass filter to smooth the raw voltage signal from the sensor before analyzing the signal to determine its minimum and maximum values. Unlike the previous two applications, we do not need a notch filter to remove power-line noise because we can add hysteresis to the comparator circuit to mitigate its effects.

Although the use of a frequency-based sensor signal complicates the system somewhat, it does have some advantages, and this is one of them.

It turns out that the undesirable signal artifacts created by bearing wear are removed rather effectively by the inherent averaging function performed by accumulating the blade counts over time. While the periods between individual blade transitions past the reference point may vary due to blade wobble, by counting the total number of transitions over a span of time, those period fluctuations tend to average out. We can improve this averaging by allowing a longer period of time for accumulating blade transitions, but we do so at the expense of system response. Of course, it's certainly possible to make the count accumulation time user configurable, probably by providing a limited offering of settings since the user will usually be unfamiliar with the nuances of filtering. Figure 8.4 shows an example of a "normal" flow signal with relatively constant times between blade transition peaks, while Figure 8.5 is an example of the signal from the same sensor after it has experienced bearing wear. Note that, although the interpeak times in the second figure vary significantly more than their counterparts in the first figure, the average inter-peak time is very similar for both signals.

At this point, the reader may be wondering why we'll be using the counter function of the dsPIC DSC timer module to gather a set of blade transition times and

Normal Steady-state Flow Signal from Turbine Sensor

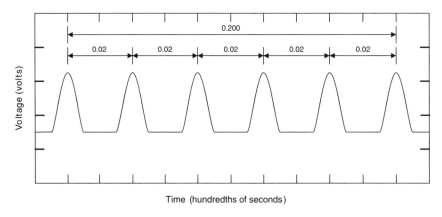

Time (hundredths of seconds)

A normal steady-state flow through a turbine with no
bearing wear should exhibit a very periodic signal

Figure 8.4. Nominal Flow Signal with No Wobble

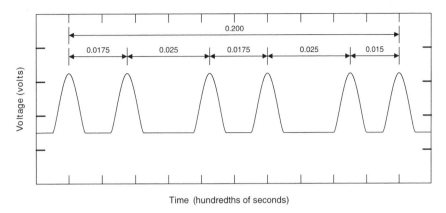

Normal Steady-state Turbine Flow Sensor Signal

If the turbine bearing is worn, a normal steady-state flow will exhibit a nonperiodic signal as the turbine blades wobble about the shaft. Although this example shows the overall period averaging out exactly after six blade transitions, the signal may not necessarily recover that quickly.

Figure 8.5. Example Flow Signal with Signs of Wobble

then compute the average transition time rather than using the dsPIC DSC's input capture features, which allow us to measure the time between individual signal transitions. The primary reason is that such a fine degree of resolution produces a great deal of jitter in the measurements, and this jitter becomes steadily worse as the turbine bearing wears. In addition, computing flow rates based on individual blade transition times tends to consume a huge portion of the available processing bandwidth at high flow rates since there is so little time between blade transitions. By gathering a large number of transitions and then averaging them, we get better results than gathering a large number of calculated flow rates and averaging those.

Data Analysis Algorithms

The data analysis algorithms for the flow sensor are somewhat more complicated than those for the temperature-monitoring and load-cell monitoring applications developed in the previous two chapters, but not much. As with those other systems, the flow-measurement system needs to check its filtered output values against upper and lower alarm limits, but periodically it also needs to read the level of the raw analog signal from the turbine sensor to determine where to adjust the biasing voltage for the peak-detection comparator circuitry. Recall that the raw signal level decreases in amplitude as the blade transition rate increases (i.e., as flow increases). Ideally, the processor should set this comparison voltage level to the same relative point in the sensor output signal even as the amplitude of the sensor output changes, so that the interpeak times don't shift simply because the sensor output level has changed. An example of the phenomenon that we would like to avoid is shown in Figure 8.6.

The algorithm for monitoring the raw sensor output and setting the comparator threshold is shown in Figure 8.7. It basically samples the sensor signal, performs some light filtering to clean up the data, and then checks whether any of the samples are outside of the previously recorded extrema values. Whenever the algorithm finds a sample whose value is either less than the current minimum value or greater than the current maximum value, it replaces the corresponding extremum with the new sample value. Once the desired number of samples have been processed, the algorithm calculates the midpoint of the minimum and maximum values detected, outputs that as the new comparator level, and then resets the recorded minimum and maximum values before acquiring and processing a new block of samples.

When determining how often to adjust the comparator level, the designer should consider the maximum rate of flow change—i.e., how quickly the flow rate can fluctuate. This is particularly important when going from a low-flow condition to a high-flow condition, because the sensor output will change from a relatively large-amplitude signal to a small-amplitude signal. If the comparator level is not adjusted quickly enough, the maximum level of the sensor output may drop below the comparator level, resulting in no comparator output transitions. This condition looks to the dsPIC DSC like a no-flow condition when, in fact, just the opposite is true! While it might at first seem that we would want to update the comparator level as fast as possible, doing so too quickly causes problems in low-flow conditions, because the update period may be shorter than the time between blade transitions. In this case, the detected extrema may not reflect the actual minimum and maximum values of the sensor output, since the update period may not be long enough to gather a full blade transition.

Effect of Comparator Level on Position and Validity of Detected Peaks

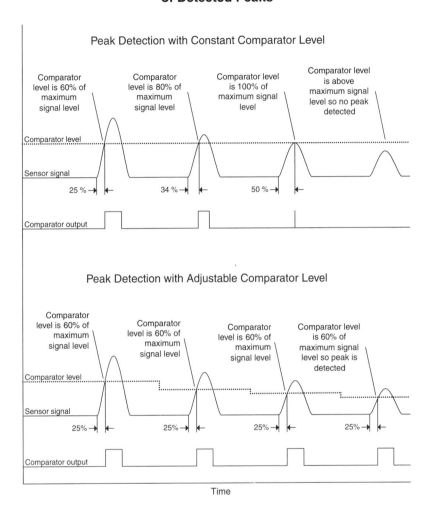

Unless the comparator level is adjusted to maintain it at a constant percentage of the sensor signal output , the distance between the leading edges of the peak detector comparator output will change with flow (since the heights of the peaks change with frequency.) The worst possible condition is one in which valid peaks are not detected, which results in reported flow values significantly lower than actual flow values.

Note that in reality, we will not know in advance the correct comparator level for the first peak that is significantly different from the preceding peak, so the first peak detection may exhibit some variance from the optimal timing. This usually is insignificant when averaged over the entire group of detected peaks.

Figure 8.6. Example of Erroneous Interpeak Period Change

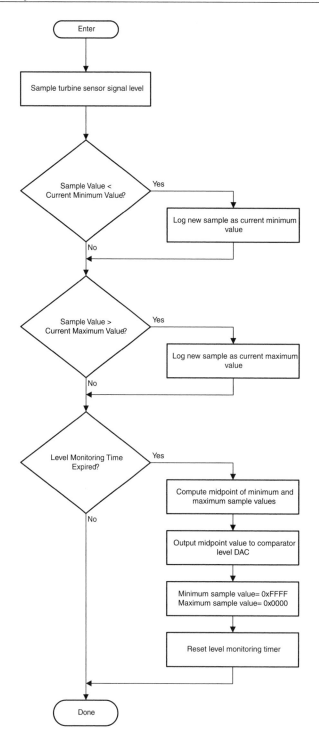

Figure 8.7. Comparator Threshold Level Computation Algorithm

Communication Protocol

The application uses the same basic communication protocol as the previous applications.

8.4   Hardware Implementation

The hardware implementation of the flow sensor is straightforward, and a block diagram of the required circuitry is shown in Figure 8.8.

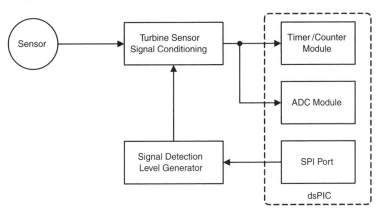

Figure 8.8. Hardware Block Diagram

Turbine Sensor Interface Circuitry

The turbine sensor interface circuitry shown in Figure 8.9 provides the sensor with power and conditions the sensor's output signal for processing by the dsPIC DSC. As with the previous applications, an instrumentation amplifier buffers the signal output by the sensor, but the gain factor is 1 since the turbine sensor produces sufficient signal amplitude on its own; the instrumentation amplifier acts as a buffer between the sensor and the rest of the system. The output of the instrumentation amplifier is filtered by a cascaded pair of second-order Butterworth filters before being fed to both the comparator that produces the digital frequency signal we'll analyze and the ADC channel that monitors the sensor signal level. Note that as with the other two examples, the schematic is for a 3.3V system, and the appropriate changes to voltage levels must be made in order to run it on the 5V dsPICDEM 1.1 GPDB.

To implement the adjustable comparator threshold, the schematic shows a Microchip 4921 digital-to-analog converter (DAC), a device that is essentially the complement to the ADC in that the DAC takes a digital input and converts that input value to an output analog voltage level. Like the ADC, the DAC is ratiometric; the output is equal to the output voltage range times the ratio of the digital value

Turbine Flow Interface Circuitry

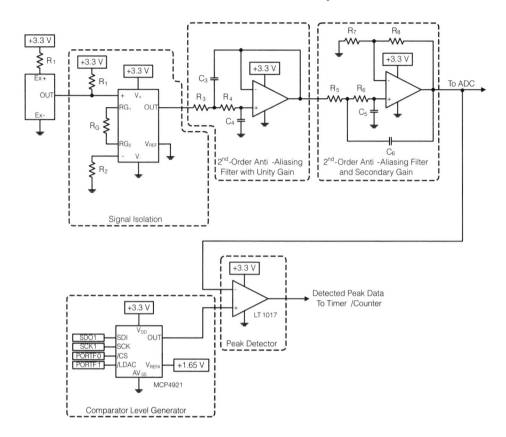

NOTES:

1. R_1 and R_2 provide an input common-mode current path and a way to determine whether the sensor is actually connected to the circuit. Resistor values should be identical and be approximately 100 KΩ to avoid loading the input signal with too low an input impedance.

2. Gain through the instrumentation amplifier is controlled by R_G according to the equation
$$Gain = 1 + (50 \ K\Omega \ / \ R_G)$$

3. Antialiasing filters have a Butterworth filter frequency response. Gain is included in the final filter stage to compensate for the instrumentation amplifier's inability to produce output voltages that are near either power supply rail. The gain in the final stage is given by the equation
$$Gain = 1 + (R_8 \ / \ R_7)$$

Figure 8.9. Turbine Sensor Interface Circuitry

written to the DAC and the maximum possible digital value. For a 12-bit DAC operating at 5V, this means that the output voltage can range from 0 to 4.9988V for input values from 0 to 0x0FFF (0 to 4095). While a 12-bit DAC works well in this application, we could easily increase the DAC's resolution by using a 14-bit (16,384 levels) or even a 16-bit (65,536 levels) device. Since the dsPIC DSC interfaces to the DAC via an SPI channel, increasing the DAC resolution would demand no additional hardware but would require a slight modification of the command data written to the DAC to support the additional data bits of the higher resolution parts. To simplify circuitry design when using the dsPICDEM 1.1 GPDB, the example software actually uses the on-board SPI digital potentiometer to set the comparator level.

Because we want to monitor the sensor signal level as it appears to the comparator, the antialiasing filter must maintain a unity gain response for frequencies in the range of interest. With its optimally flat gain response through the passband, the Butterworth filter is ideal for this purpose, and by ensuring that the stopband begins well beyond the highest valid frequency content of the sensor signal, we can effectively remove any undesired frequency components without adversely affecting the relevant spectral content of the sensor signal.

8.5 Firmware Implementation

The flow-measurement firmware uses basically the same structure as the two previous examples. The main components are:

1. the data acquisition module consisting of the blade transition accumulation routine, the ADC sample timer interrupt-service routine, and the ADC interrupt-service routine,

2. the data filtering module,

3. the data analysis module, consisting of the blade-count analysis and comparator level-adjustment routines,

4. the hardware error-detection module, and

5. the communication module.

Data-acquisition Module

The data-acquisition module is responsible for counting the blade transitions in a given time period and for adjusting the comparator threshold signal level to optimally detect those transitions. To do the latter, the dsPIC DSC samples the sensor signal level going to the comparator, determines an appropriate comparator threshold voltage, and outputs that value via the SPI to the digital potentiometer that generates the analog threshold voltage.

Sensor Signal Level Monitor

As in the temperature and load-cell sensing system, the application employs Timer 3 to create the 1-kHz ADC sample clock. Because we're just trying to find the extremes of the sensor signal during a given time period, the ADC is configured to acquire 8 samples before interrupting. The flow sensor data is unipolar, so we use unsigned fractional mode for the data. When the ADC interrupt is called, the ISR simply unloads the sample buffer into the g_frSensorSignal filter buffer and sets the EVT_FILTER event, which causes the main processing loop to filter the sensor level data and update the DAC's output voltage.

Blade Transition Counter

The dsPIC DSC's timer module can be configured to count the number of rising edges on a timer's external clock input pin, which in this case will be connected to the output of the comparator. By using a second timer to generate an accurate time base, we can determine the number of blade transitions that occurred during a given time period and thus determine the corresponding flow level.

This application uses Timer 5 to create a 16-bit synchronous counter, allowing us to measure blade transitions that are spaced fairly far apart and extending the lower end of the flow levels we can measure. The data-acquisition module also uses Timer 3 to generate the 500-ms timebase used to accumulate the blade counts, simply reading and resetting the accumulated Timer 5 counter value every 500 times through the 1 msec Timer 3 interrupt. For the flow sensors with which the author has experience (4- and 6-blade models), a blade count accumulation time of 500 ms seems to work well, but the optimal time period depends upon the responsiveness required by the specific application as well as the anticipated flow rates. An accumulation time of 500 ms produces an update rate for the computed flow rate of 2 updates per second (1/500 ms per update), which may or may not be appropriate for a given situation. If greater responsiveness is required, the accumulation time can be decreased (thus updating the threshold more often), but doing so decreases the inherent averaging properties of the algorithm and also raises the minimum detectable flow rate, since the system must see at least one blade transition per accumulation period in order to see a nonzero flow.

Upon expiration of the 500 ms timer, the Timer 3 ISR logs the current 16-bit value of Timer 5 (which contains the blade transition count) to the global variable `g_ui16SensorCount` and sets the `EVT_ANALYZE` event to signal the main processing loop that it needs to invoke the data analysis routine for the flow rate.

Data Filtering Module

Unlike the previous two examples, in which the sensor signal was filtered to improve the parameter analysis, the filtering in this application is performed to obtain a better reading of the signal's voltage level so it can be used to set the comparator level optimally.

Sensor Signal Level Filtering

The sensor signal level filter is a low-pass filter. Unlike the filtering scheme used in the load-sensing applications, this system does not attempt to apply a notch filter to remove power-line noise that's been coupled into the sensor signal, because signals with that frequency are completely valid for this application. Removal of even a narrow band of frequencies around 50 Hz and/or 60 Hz would result in a "deadband" of valid flow levels within which the comparator output level would not be updated. Since the effect of power-line noise on the comparator signal is minimal, we can safely ignore noise coupled from the power-line in this situation.

The filter uses a sampling rate of 1 kHz, which won't be much of a burden because we're only going to be monitoring one channel and don't have to process the digitized data on a sample-by-sample basis. Minimizing the processing overhead required for digitization allows us to devote more time to filtering, so we can implement a filter that has an extremely flat gain through the passband, as shown in Figure 8.11. This filter is a 59-tap minimum 4-term cosine window and, as the reader will note, it offers unity gain through the passband.

Data-analysis Module

The data-analysis module is responsible for two basic tasks:

1. checking the latest filtered-flow rate value against the user-specified alarm limits and setting the display states of the corresponding alarm limit LEDs appropriately,

2. analyzing the buffered sensor output signal level to determine

a. whether the sensor is actually connected to the system, and

b. if the sensor is attached, computing the optimal comparator threshold voltage and outputting that value to the DAC.

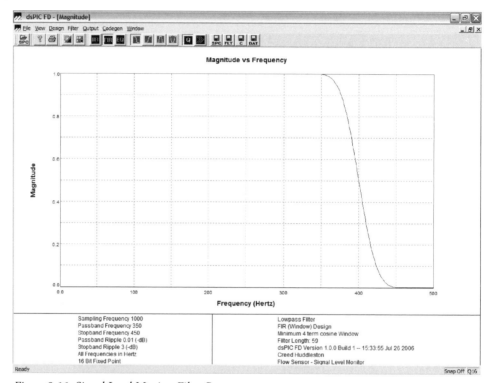

Figure 8.11. Signal Level Monitor Filter Response

Flow-rate Analysis

The data analysis of the flow-rate value is the same as the two previous applications; it simply checks the latest filtered flow-rate value against the specified alarm limits, sets the state of the corresponding limit LEDs appropriately, and then increments a global counter that tracks how much time has elapsed since the previous flow rate value was reported via the communication port. If a second has passed since the previous report, the function sets the EVT_REPORT_RESULTS event to alert the main processing loop that it should report the latest flow rate value via the communication port. Finally, the routine clears the EVT_ANALYZE_DATA event to show that the data analysis has been performed for the current filtered flow rate and returns.

Signal-level Analysis

The sensor signal-level analysis first verifies that the sensor is indeed connected to the system input. If the sensor is not connected properly, the pull-up resistor on the input to the high (positive) side of the instrumentation amplifier will force the instrumentation amp to a constant output near the positive voltage rail. In all other cases, the signal will be below this rail, although it may temporarily approach it when the turbine blade is passing the illumination point (since that is the time of maximum reflected signal from the blade). To confirm that the sensor is attached, the application simply examines the filtered sensor signal level and declares that the sensor is not connected if that signal level exceeds a preset value for longer than, say, 0.5 seconds. The precise minimum detection voltage level and the time period required to trigger a hardware-error detection can be adjusted by the designer, but the combination should be somewhat beyond the maximum signal level and level presence time that would occur during valid operation.

If the preceding analysis indicates that the sensor is not connected to the system, it sets the global `g_bHWError` flag to indicate that fact, lights the Hardware Error LED, and then skips the comparator threshold analysis since the sensor signal would be invalid anyway. If the sensor is connected, however, the routine clears `g_bHWError`, turns off the Hardware Error LED, and proceeds with the comparator threshold analysis.

The comparator threshold analysis is extremely straightforward; it examines the filtered signal level values on a sample-by-sample basis to determine the minimum and maximum values over the preset threshold update period, computes the midpoint of the two extrema, and outputs that midpoint value to the DAC to set the new comparator level.

Once the comparator threshold analysis has completed (or, if the sensor is missing, immediately after the sensor-detection analysis has finished), the main processing loop clears the `EVT_ANALYZE` event.

Communication-protocol Module

Whenever the main processing loop receives an `EVT_REPORT_RESULTS` event, it calls the `FormatResultsMsg()` routine. This function formats the latest flow data, then the main loop transmits the formatted message to the host system using the routine `TransmitResults()`. Once the result message has been sent, the main processing loop resets the `EVT_REPORT_RESULTS` event to show that the event has been processed. The protocol used is the same as in the temperature and load-cell monitoring systems, so it will not be discussed further here.

8.6 **Summary**

The flow-meter application developed in this chapter introduced us to a form of data acquisition in which the frequency and not the voltage level of the sensor output signal contains the parameter data of interest. This is a simple, effective way to encode sensor data, but there are caveats regarding its use in other applications. Care must be taken to ensure that the system can process the minimum and maximum valid signal frequencies, with special attention being devoted to very high-frequency rate signals, since those can easily swamp a poorly designed system's ability to process. Obviously, we could improve the application's performance through enhancements such as flow-rate sensitive blade count accumulation (allowing longer accumulation time for low-flow conditions to extend the lower range of measurement and shorter accumulation times for high-flow conditions to increase responsiveness) or by adding a current sense resistor to the sensor's power line (to augment our diagnostic abilities), but the focus of this chapter has been to introduce the reader to a general sensor interface approach rather than to develop an exhaustive list of features.

In the final chapter, we'll look at where the intelligent sensor market is headed from both technical and business perspectives.

Endnotes

1. Transducer Interfacing Handbook: A Guide to Analog Signal Conditioning, Edited by Daniel H. Sheingold. Copyright 1980, 1981 by Analog Devices, Inc., p. 24.

2. The RotoFlow™ line from CITO Products, Inc.

Where Are We Headed?

The best way to predict the future is to invent it.
—Alan Kay[1]

Now that we're reasonably comfortable with intelligent sensors and the concepts required to implement them, let's turn our attention to what the future holds for these powerful devices. As with any new technology, today's intelligent sensors have tremendous potential, but they also have a number of issues that must be addressed in order for their use to become pervasive. How can we expect the field to behave in the coming years? What are the drivers for intelligent-sensor diffusion into the marketplace, and what are the constraints upon it? Into which markets can we expect to see significant intelligent sensor penetration, and what new capabilities are reasonable to expect? These are the questions we'll explore in this chapter. While we may not be able to answer them all, the reader should leave with at least an understanding of the signposts that mark important points in the intelligent sensor's evolutionary path and provide the reader with resources that will allow him to delve more deeply into this fascinating subject.

9.1 Technology Trends

Intelligent sensors consist of three basic building blocks—sensing elements, computational elements, and communication interfaces—and the underlying technologies of all three building blocks are evolving at a rapid pace. In the next few years, we will witness an explosion in the capabilities of intelligent sensors, an explosion that will improve functionality and connectivity significantly while simultaneously reducing system cost. Let's look more closely at the developments that are occurring in each of these three realms.

Sensing-element Trends

One might suppose the field of transducer development to be fairly static, but that supposition would be totally incorrect. Advances in this area have significantly increased the sensitivity of a number of different types of sensors, particularly in the area of chemical analysis. With this increased sensitivity, designers can create systems that are physically smaller, that consume less power, and that require less of the substance being analyzed in order to obtain an accurate reading.

Nowhere is this trend more pronounced than in the world of microfluidics, which is devoted to controlling and/or analyzing extremely small quantities of a fluid. In this case, small is very small, on the order of microliters or even nanoliters.[2] With its concepts being applied to everything from inkjet printers to lab-on-a-chip devices that scrutinize blood, the field of microfluidics covers a lot of ground. A subcategory, known as *digital microfluidics*, is concerned with the precise manipulation of fractions of a drop in order to assay the fluid's chemical makeup or to perform some other useful task.

As one might expect, dealing reliably with such small quantities of a substance depends upon a system's ability to accurately measure important parameters associated with the droplet. In particular, it turns out that the ability to precisely measure and control the pressure and temperature of the substance and the pathways it must traverse is critical; failure to maintain the proper conditions will bring a microfluidics system to its knees. If, however, one is able to sustain the required environment, the rewards are enormous: a reduction by several orders of magnitude in the amount of a substance that must be tested, reporting of the results in a matter of minutes rather than a matter of days, increased system portability, and reduced cost. Although not the only area of significant sensor technology advancement, the field of microfluidics is certainly one of the more obvious ones.

Designers can expect to see this trend of miniaturization and increased transducer sensitivity to continue into the foreseeable future. Advances in complementary fields such as nanotechnology will expand the environmental conditions in which particular sensors can be deployed as well as creating entirely new types of sensors. In order to effectively employ these advances, however, designers must not only stay abreast of new developments in transducer technology; they must also learn about the underlying physical phenomena from which the transducers derive their functionality.

Computational Element Trends

By now, most of us have heard about Moore's Law, which is popularly taken to be that computing power doubles about every two years.[3] While we can certainly expect to see computational horsepower increase in the future, embedded microprocessor systems such as the ones that we've been exploring have other considerations besides raw computational throughput. Power consumption, physical size, and open hardware and software architectures are also important factors that, in certain applications, may be more important than processing power.

Dramatically Lower Power Consumption

In many cases, the power consumption of the system is a critical factor prolonging battery life in portable or remote applications and in reducing undesirable self-heating. Many digital electronics, including the dsPIC DSC and most processors, are constructed in such a way that their power consumption and the heat they produce are proportional to the speed at which they are clocked. Double the clock speed to run the application twice as quickly, and you also double the amount of power required by the chip. Although that's often a problem, it also offers a potential solution; by shutting down portions of the chip when they're not needed and/or by slowing down the clock under those conditions, we can dramatically reduce power consumption. This is particularly important for single-use battery-operated systems that may have to perform for weeks, months, or even years on a single set of batteries.

We are already seeing drastic reductions in the amount of power required to run both processors and the circuitry required to condition the sensor signal, and we can expect to see this trend accelerate in the future as the demand for much longer battery life in mobile devices (measured in years, not days) continues unabated. One particularly flamboyant advertising campaign by Microchip already employs a microprocessor-based temperature-measurement system powered by a grapefruit to demonstrate just how frugal their components are when it comes to power consumption,[4] and such low-power operation will become the norm, not the exception, in the future.

Unfortunately, nothing's free in this life; in order to cut power consumption significantly, chip designers have had to reduce the power-supply voltages to 3.3V and often less. That has serious repercussions, particularly in the analog signal chain. By lowering the power-supply voltages, we effectively increase the noise level in the analog signal chain relative to the power voltage span. For instance, if we have

10 mV of noise in an analog signal that has a 5V span, we've got a noise level of 2% (10 mV / 5V = 0.02). That same 10 mV of noise in an analog signal with a 2.5V span has essentially doubled to 4%, which can cause some pretty serious errors if not handled properly. This effect may be exacerbated in excitation voltages that have to be run a significant distance in an electrically noisy environment; indeed, one of the reasons that many older sensor systems used ±10V or even ±15V as their sensor excitation voltage is that such a large span minimizes the effect of noise on the output signal.

The other negative aspect of reduced voltage parts is that they are more difficult to interface to legacy 5V or higher systems. While not impossible by any means, connecting newer low-voltage systems to older higher-voltage systems usually requires some sort of level translation for reliable operation. Although it's not unusual for I/O signals on newer parts to be 5V-tolerant, the output signals on the lower-voltage parts may not be able to actually drive the 5V system's inputs at a high enough voltage level. This problem will fade in time as more and more systems migrate to lower-voltage power requirements, but it is a factor that must be accounted for in designs in the near term.

Continual Size Reduction

Hand in hand with the reduction in the power requirements for intelligent sensor systems will be the drive to further compact the size of these systems. Small, battery-powered sensor systems the size of an American quarter are already available, but future systems will be even tinier. Known as *motes*, these completely self-contained devices have on-board radio links that allow them to form ad hoc communication networks (more on this in the section Communication Trends) for the exchange of data between other motes in the network or host systems. As impressive as this may sound, motes have been developed that are about 5 mm to a side, and the goal is to eventually reach systems that are 1 mm^3 in volume[5]. That tiny package, about the size of a grain of sand, will hold everything: sensor, processor, radio, and battery.

Such miniaturization brings with it a host of packaging and operational issues, particularly finding a way to create an energy source in such a small volume that has sufficient energy to run the system for an appreciable length of time. Communicating with such a small device is another important issue, since one would be hard-pressed to connect to something that small with standard cables. We'll now turn to the issue of communication, both with motes and with larger systems, in the next section.

Communication Trends

As important as the advances in electronic circuitry will be to the future of intelligent sensors, it will be the incredible enhancement of communication technology that will drive dramatic growth in intelligent-sensor usage. Particularly for motes and other very small sensor systems, this communication will be handled wirelessly through ad hoc sensor networks that automatically form with little or no user intervention, allowing rapid deployment. Physically larger sensor systems may still have wired communication connections, but they will increasingly be based on high-speed (100 Mbps and faster) networking standards to facilitate the aggregation, dissemination, and storage of the massive quantities of information these systems generate. Let's look at some of the specific ways intelligent-sensor communication will evolve in the future.

The Pervasive Internet

Harbor Research, Inc., a strategic consulting and research firm whose focus includes the intelligent sensor sector, has coined the phase the "Pervasive Internet" to describe the convergence of pervasive computing and widespread networking of devices connected to the Internet.[6] Based on current trends, this is an apt description; the ubiquitousness of devices of all types (cell phones, PDAs, laptop computers, etc.) that support Internet access has mushroomed. The connection of intelligent sensor systems to the Internet is starting up that same trajectory as well. While some may consider this approach to be a devolution into "gadgetry," the truth is that there are many applications that would benefit from the control such communication would offer. For instance, it's not at all hard to imagine a car whose engine parameters can be set by the user; extra power could be made available when the family goes on vacation and needs to tow a boat or the top speed might be limited for a new student driver. Although not commercially available today, such a configuration is technologically feasible; it only awaits those astute enough to offer the service.

Before such a connected system becomes a reality, however, there are some serious issues that must be addressed. The first such issue is rather prosaic: finding addresses for all of those devices. The current Internet protocol, known as IPv4, supports up to 4.3×10^9 unique addresses (4.3 billion), which sounds like quite a few until one realizes that the address space works out to less than one address per person on the planet. Given that each person is responsible for numerous items in their personal and professional lives (automobiles, appliances, computers, and nonelectronic devices) and given that organizations (schools, corporations, government, etc.) are stewards of many, many more, the current addressing scheme is completely inadequate to the task of connecting these items together. Fortunately, the next generation of the Internet Protocol, IPv6, supports far more addresses, 3.4×10^{38} to be precise. That's easily enough for our needs well into the foreseeable future.[7] Other approaches are available as well, but none have progressed as far as IPv6.

A second issue is the matter of bandwidth; how do we efficiently transfer the vast quantities of information that would be generated by such an interconnected system? Although we've all seen the rapid growth of broadband Internet links (cable, DSL, etc.), communications on the scale envisioned will require significantly greater transmission capacity than is currently available. The Internet2 project[8] is developing a much faster Internet infrastructure that runs at 10 Gbps (10^9 bits per second), but even this is only a start, and as of this writing (fall of 2006) is only available in the United States. Significant work remains to be done in the area of high-speed routing and information selection to reduce unnecessary network traffic to levels that can be supported reliably and cost-effectively.

Wireless Communications

If all of the newly connected systems required cabling to communicate, the situation would be hopeless; it would be cost prohibitive and too unwieldy to be feasible. Fortunately, wireless communications are becoming widespread, easily implemented, and very cost effective, allowing systems to "cut the cord" in many situations. Depending upon the application, this may refer to eliminating only the communication cabling, but increasingly it also means using battery-powered or passively powered devices to do away with the power cabling as well.

A number of different wireless communication schemes are available that are tailored to specific classes of applications. The relatively new ZigBee™ protocol is particularly suited to low-power, low-data rate applications that may need to run for years on a single battery. In these applications, communications are limited in both duration and frequency of occurrence in order to reduce power consumption. Other protocols, such as the popular 802.11x protocols used for wireless Local Area Networks (LANs), are capable of continuous communication and are used for applications such as real-time monitoring and control (with provisions for temporary interruptions, of course). Both ZigBee and 802.11x protocols are intended for geographically limited areas. At the other extreme are wireless schemes that employ cellular modems or other techniques to connect to devices across the globe. We can expect to see much broader use of wireless technologies to reduce infrastructure costs, to simplify installation, and to improve usability.

Security

A tremendous concern going forward is how to secure networks of connected devices against unauthorized monitoring or worse—tampering. Depending upon the type of information being reported by the system, this concern is no less valid for intelligent sensing systems than it is for financial institutions. In particular, sensitive information that is being used to control processes must be protected against both intentional corruption and "snooping" to prevent outside personnel from damaging the process or from gathering proprietary information. The need to ensure security applies to both wired and wireless networks.

Protocol designers now recognize the importance of security to robust communications, and recently developed protocols such as ZigBee incorporate security as a foundational element in their design rather than as something added as an afterthought. Unfortunately, added security means added overhead to the information bit stream, reducing data throughput and, depending upon the implementation, increasing the hardware and/or software cost of the system. In the end, however, the potential cost of a breach of security generally outweighs the actual costs incurred to include it in the product.

Ad Hoc Networking

In many networking applications, one of the greatest expenses in terms of time and money is the administration of the network. Maintaining a network of any significant size usually requires the attention (and cost) of a highly skilled network administrator, something that many of the applications targeted for intelligent sensors simply cannot afford. What is required in order for sensor networks to thrive is the ability for a relatively unskilled individual to quickly deploy these networks, and for the network nodes themselves to automatically configure themselves for proper operation. Such networks are known as *ad hoc* or *self-organizing networks*, and we are seeing the first reliable systems emerge. ZigBee is an example of such a networking scheme, allowing nodes to enter and leave the network without requiring a massive reconfiguration, and there are other implementations as well, though they are often proprietary. We can expect to see significant advances in this area in the relatively near future (the next few years).

Having looked at the technological trends shaping the intelligent-sensor space, let's now look at the economic trends that will shape its acceptance.

9.2   Economic Trends

Three key economic developments will drive the future adoption of intelligent sensors: the demographics of an aging population, the increasing globalization of operations, and the creation of new business opportunities. All three trends point to strongly increased demand for intelligent-sensing systems.

Demographics of an Aging Population

In all developed countries, the birth rates have fallen below that required to sustain their populations, and this has been the case for some time. The resulting aging of the population, and in particular of the labor pool, has been recognized for quite a while, but its effects are just now starting to make themselves felt with the retirement of the baby-boom generation in the United States and their post-World War II cohorts around the globe. In industry, medicine, and other activities, we will shortly lose significant portions of the workforce, with no adequate replacements immediately in sight.[9] If we are to maintain the economic engines that provide our current level of goods and services, it follows that per-worker productivity will need to increase dramatically in order to replace the productivity lost with retiring workers. The most effective, indeed perhaps the only, way to accomplish that goal is through the use of intelligent sensors that self-organize, self-diagnose, and that on the whole require far less human interaction than current systems.

By the way, this is not just a developed-nation problem; the trend toward reduced birth rate is true throughout most of the world, including the developing nations. Unless reversed (something that could obviously happen, though it would take time), all countries will eventually face the same dilemma.

Increasing Globalization of Operations

No matter where you stand on the issue of globalization, it is a fact of life, and it will continue for the foreseeable future as companies deploy resources to geographic locations in which they deem the resources are employed most efficiently. While globalization may reduce costs for the organizations, it also makes it more difficult for those tasked with managing dispersed groups to gather and to act upon the information they require in order to make effective decisions. To fill this gap, companies will increasingly make use of intelligent-sensing systems to provide real-time information to key decision makers and to technical personnel responsible for the smooth operation of the organization. Because these decision makers and technical personnel may not be—indeed often are not—physically located with the activities for which they are responsible, the connected nature of intelligent-sensing systems is critical to their ability to do their jobs well.

As with the issue of an aging population, globalization is not just a concern for fully developed countries. Countries such as India and China, which are typically perceived as the recipients of outsourced operations, now find that they themselves must outsource work to distant areas within their own country or even to other countries in order to remain cost competitive.

New Business Opportunities

Just as the network of railroads in the western United States in the latter part of the nineteenth century offered enterprising organizations new business opportunities by drastically reducing the time and cost to transport goods and people, so too will the evolving field of intelligent sensors provide new ways for insightful businesses to reap the benefits of pervasive computing power and globally networked devices. Already, leading-edge companies are employing intelligent sensors in a wide variety of applications to deliver value to customers and to recover a portion of that value in the form of revenues. In an interesting one-page display, Harbor Research identifies nearly a hundred application segments that can benefit from networked intelligent devices.[10]

With intelligent-sensing systems, sensor providers and integrators now have the ability to create on-going revenue streams from services related to monitoring and using those sensors, rather than having to rely on the traditional unit-sales of individual sensors. As mentioned in Chapter 1, this can provide tremendous benefits to the end users while providing steady revenue to the supplier and strengthening the relationship between the vendor and buyer. In a world increasingly dominated by commodity-purchasing approaches by buyers, this aspect gives vendors a way to differentiate their offerings and to avoid the perception by customers that the vendor's products and services are just like everyone else's.

9.3   Summary

The basic premises of this book are that intelligent sensors are the future of the huge and growing sensor market and that the dsPIC DSC provides an excellent foundation upon which to build a wide variety of intelligent sensors. What has perhaps not come through as clearly is that the functionality of these sensors is limited primarily by the imagination of the designers who create them. Although the field of intelligent sensors requires broad knowledge in a number of areas (analog signal conditioning, digital signal processing, and to a lesser extent digital design[11]), it affords a canvas upon which a creative designer can paint some incredible applications.

Ultimately, as Alan Kay notes in the chapter's opening quotation, "the best way to predict the future is to invent it." The reader is encouraged to delve into this realm more deeply and to bring his or her unique knowledge to bear on a particular application or group of applications. Not only is the design of intelligent sensors a wonderful intellectual challenge, each new sensor adds to our ability to understand and to shape our world. It may be financially rewarding, as in the case of a system that's widely used to manufacture billions of parts worldwide, or it may change a person's life by making the world more accessible through the creation of a smart prosthetic. If you apply your acumen and insight along with the information presented in this book, you can make a significant difference with your implementation of intelligent sensors.

Endnotes

1. Kay is a giant in the field of computer science who was one of the founders of Xerox's famed Palo Alto Research Center (PARC). A pioneer in the development of modern object-oriented programming and windowed graphical user interfaces (GUIs) with his work on the computer language Smalltalk, Kay designed a graphical, object-oriented personal computer during his Ph.D studies at the University of Utah. While that may not seem particularly impressive today, given that Kay did it in 1969, the feat is seen as truly groundbreaking. The man knows whereof he speaks.

2. A *microliter* is one-millionth of a liter (10^{-6} liters), and a *nanoliter* is one-billionth of a liter (10^{-9} liters). To put that in perspective, those volumes are less than a single drop of fluid.

3. Moore's Law is named for Gordon Moore, a cofounder of the Intel Corporation of PC processor fame, who made the prediction in 1965. In fact, according to the Intel website, Moore actually stated that *the number of transistors on a chip* doubles approximately every two years. Given that the number of transistors in a microprocessor offers a rough approximation to the chip's processing power, the "law" evolved over the years into its more popularly known form. An in-depth look at Moore's Law can be found at http://www.intel.com/technology/silicon/mooreslaw.

4. The demonstration uses a Microchip PIC microcontroller fabricated using their nanoWatt™ technology.

5. Kristofer Pister, professor of electrical engineering at the University of California at Berkeley, as quoted in the March 23, 2004 online version of ComputerWorld magazine, which can be found at http://www.computerworld.com/mobiletopics/mobile/story/0,10801,79572,00.html.

6. The Harbor Research website (www.harborresearch.com) has a number of insightful white papers on business opportunities in the Pervasive Internet space, as well as on the specific strategies appropriate to capitalize on those opportunities.

7. To put the number of IPv6 addresses in perspective, the new protocol allows roughly 50 octillion (50×10^{27}) addresses for each of the approximately 6.5 billion people currently on Earth. Interestingly, IPv6 also sprang from the fertile minds at PARC (see footnote 1).

8. More information on the Internet2 project is available at the organization's website, www.internet2.edu.

9. That's not to imply that younger workers are incompetent, merely that the number of younger workers is insufficient to replace those retiring.

10. *Venue Segmentation Map for Intelligent Device Networking and Management*, available for download from www.harborresearch.com.

11. This statement is in no way meant to denigrate the importance of solid digital design; it is instead an acknowledgement of the fact that so much functionality that previously would have required separate digital circuitry is now integrated into the dsPIC chip itself. Effective digital design techniques are still required to use the dsPIC DSC, but the degree of system integration reduces the demands on that skill.

Software on the Included CD-ROM

The software on the included CD consists of an on-disk website with links to valuable resources on the Internet and the source code and project files for the three applications developed in the book. To view the website, either use Windows Explorer™ (the file management program, not to be confused with Internet Explorer, which is a web browser) to find the file index.htm in the root directory of the CD and double-click on the file. That should start your web browser and load the first page of the site. Alternatively, you can enter:

```
D:\index.htm
```

in the address bar of your web browser and press the "Go" button in the browser to load the first page. Note that this assumes that your CD drive is drive D; if this is not the case, simply substitute the appropriate letter in the path.

A.1   On-disk Website of Resources

The links on the website provide the reader with an easy way to obtain some of the information discussed in the book—in particular, access to reference materials and vendors that provide useful components or equipment. As with any site that references pages in another website, some links may become stale and no longer work. If that happens, search on the Internet using the phrase used for the link (as opposed to the link itself), and you should be able to find the appropriate data if it's still available.

Please note that the links are included to help the reader find information quickly and are not meant as an endorsement of any particular product or vendor by Elsevier Science and Technology Books or Newnes (with the obvious exception of their own books).

A.2   Source Code for the Three Applications

The complete source code for all three applications, including project files for use with the Microchip C compiler, are included in separate directories:

- \Source\Thermocouple – the thermocouple sensor system

- \Source\Load Cell – the load-cell sensor system

- \Source\Flow Meter – the flow-meter sensor system

All of these applications require the Microchip C compiler, v2.02 running under the Microchip MPLAB v7.40 Integrated Development Environment (IDE) in order to compile and link (later versions should work as well, but earlier versions may not). The MPLAB IDE is available for free from the Microchip website (www. microchip.com), and a student version of the C compiler is available for free from the site as well.

If you would like to alter the digital filters used in any of the applications, you will need to purchase a copy of the dsPIC Filter Design™ software, which is also available from the Microchip website. In order to run the applications, you will need a Microchip dsPICDEM 1.1 General Purpose Development Board and an ICD 2 in-circuit debugger to download the program file to the dsPICDEM board. Both of these items can be obtained through the Microchip website as well.

B

Initialization of the dsPIC DSC and the System Start-up Code

When the dsPIC DSC first boots out of reset, it begins running the code residing in the Reset vector location (00000H) with all interrupts disabled. In order to operate properly, the code must jump past the Interrupt Vector table and the Alternate Vector table to the application code space and then start running what is known as the *start-up code*. A critical component of the firmware, the start-up code configures the C runtime environment so that the application itself (written mostly, if not entirely, in C) can execute as expected. This step must be taken before any calls to C code, or the C code may fail, inevitably in some manner that is difficult to diagnose.

Because the start-up code initializes the C runtime environment, it is specific to the compiler being used and is distributed as part of the compiler. For the Microchip compiler, two start-up code files are provided, one (crt0.o) that initializes all initialized data to values that are read from the program memory and that clears all uninitialized data to 0, and a second (crt1.o) that does not initialize any data. Other than that, the two modules are the same and perform the following actions:[1]

1. Initialize the Stack Pointer register (W15) and the Stack Pointer Limit register (SPLIM) using values generated by the linker.

2. If a .const section is defined, the start-up code maps it into the program-space visibility window by configuring the PSVPAG and CORCON registers appropriately.

3. Perform data initialization (crt0.o only).

4. Call the main() application entry point to start the user's application.

Note that the standard start-up code included with the Microchip C compiler does not perform any hardware initialization other than the registers mentioned above; the user's application code is responsible for configuring the hardware and any required other software set-up.

Most of the time, the start-up code is transparent to the application programmer. Included as part of the standard linker files are instructions to place a GOTO statement to the start-up code at the Reset vector location and further instructions to link in the standard start-up code module `crt0.o`, which performs the required C system initialization and then calls the user's program. Occasionally, however, the application may require immediate configuration of some critical system resource, in which case the user can modify the assembly-language source file `crt0.s` or `crt1.s` to implement the appropriate changes. If such a modification is required, the programmer must include the file `crtx.s` (where x is either 0 or 1, as necessary) in the list of project source files to ensure that it is properly linked into the final application.

Endnotes

1. This information can be found in the section, Startup and Initialization, of the *MPLAB C30 C Compiler User's Guide*.

C

Buffered, Interrupt-driven Serial I/O

In many discussions of serial communications, the author assumes that the reader has access to routines that reliably transmit and receive data so the subject is addressed at a very high level. Although often treated as an exercise left to the reader, the topic of developing a dependable low-level communication interface is one whose solution is often far from obvious and changes from processor to processor, even when discussing processors produced by the same manufacturer. The key to solid serial communications in real-world conditions is to create a buffered, interrupt-driven I/O framework. Let's examine what we mean by that mouthful of technical jargon.

An I/O *framework* is simply an architecture that we use to perform all of our input ("I") and output ("O") serial communications. By adhering to the framework, we should be able to ensure that our device receives and transmits all of the data that it needs to handle.

An *interrupt-driven framework* is one in which we allow the communication hardware to interrupt the processor so that the processor can deal with the new communication condition immediately and then return to what it was doing. If we were to use the alternative, a polling framework in which the processor periodically polls (queries) the communication hardware to see if anything needs to be done, we would have to design the application firmware to guarantee that the processor finished all of its other tasks in time to check for the fastest possible communication event. Given that communication events can come fast and furiously, a polled architecture can be severely restrictive for all but the slowest communication channels.

Finally, a *buffered, interrupt-driven framework* is one in which the data to be transferred between the application and the communication channel is stored in queues, with the underlying interrupt-service routines and the application program managing access to the queues in such a way as to ensure that the queue data is *coherent* (accurate, with no conflicts between access by the ISRs or the application). This approach offers maximum flexibility for the application while ensuring highly reliable communications with minimum processing overhead.

The heart of this type of framework is the data queue, which in practical terms consists of a receive data queue to hold data received previously by the device from other systems and a transmit data queue that holds data to be sent by the device to other systems. Obviously, the amount of data that the device may be asked to handle over time will far exceed the limited memory of any real-world processor, so the queues are designed as *circular buffers*, with data being added to the buffer in a sequential manner until the end of the buffer is reached, at which point new data is added to the buffer starting at the beginning. Data is removed from the buffer in a similar manner. Conventionally, the point at which data is being added to the buffer is known as the *head* of the buffer, while the point from which data is being read from the buffer is known as the *tail*.

To use the queues, functions at the application level and those of the interrupt-service routines maintain two sets of indices to determine the buffer locations that the corresponding functions will use next. The specific purpose of each index depends upon the execution level (application or interrupt) of the function using the index and the queue upon which the function is operating. For instance, when the application wants to send data to another system, the application-level function *writes* that data to the transmit data queue and updates the index the application maintains of the next byte in the transmit data queue to which it can write. When the interrupt service routine that handles data transmission notes that there is new data to send, it will *read* the data written by the application level, update its (the ISR's) index of the next byte to read from the queue, and then send the data to the communication hardware. In this case, the application-level functions are responsible for maintaining the head of the circular transmit data buffer, while the ISR functions are responsible for maintaining the tail of the buffer. The situation is reversed for the receive data queue, with the ISR being responsible for the head of the buffer and the application-level code handling the tail.

This approach works well until something happens that prevents one or both of the queues from emptying as quickly as they are filled. In this case, the head eventually overtakes the tail, a condition known as *buffer overflow*. If the buffer is allowed to overflow, all the data between the tail and the now rampaging head is lost. Since some data is going to be lost under these conditions no matter what, this implementation chooses to ignore data received after the buffer is full, rather than to lose data that has already been received. Note that once the buffer empties by one or more characters, data reception will continue again.

C.1 Pseudo-code for the Framework

The actual code for the serial interface is contained in the files `CommIF.c` and `CommIFDef.h`. The pseudo-code here provides a high-level reference to the tasks being performed by the interface. To use the interface, the application must first initialize the communication hardware and associated global state variables and enable the corresponding processor interrupts. The application can then read data from and write data to the serial ports using the functions shown below.

C.2 System Initialization

To initialize the communication system, the user calls the function CommInit() with parameters specifying which UART (1 or 2) to use, along with the requisite communication parameters (bit rate, parity type, and number of stop bits).

Calling sequence:

```
Uint16 CommInit(Uint8 ui8Port,

                Uint16 ui16BaudRate,

                Uint16 ui16Parity,

                Uint16 ui16StopBits)
```

Example: Initialize UART 1 to communicate at 19.2 Kbps, using no parity and 1 stop bit

```
CommInit(UART_1, 19200, PARITY_NONE, STOP_BITS_1);
```

C.3 Reading Data From the Interface

To read data from the communication interface, the application calls the routine `CommGetRxPending()` to determine whether there is any data available to read, and if the return value (which indicates the number of pending data bytes) is greater than 0, the program reads in the next byte of data from the receive data queue by calling the function `CommGetRxChar()` with a parameter that is a pointer to an 8-bit location to hold the data read from the queue. If `CommGetRxChar()` returns a nonzero status value, an error has occurred (see the file `StatusDef.h` for a complete list of status code values), and the application should not use the data returned in the buffer pointed to by the parameter.

Calling sequence:

```
Uint16 CommGetRxPendingCount(void);

Uint16 CommGetRxChar(Uint8 *pui8Data);
```

Example: Check for pending data from the communication port and read it in.

```
Uint8 ui8Data;       // Buffer to hold data from Rx queue


if (CommGetRxPendingCount() > 0)

    CommGetRxChar(&ui8Data);
```

C.4   Writing Data to the Interface

To write data to the communication interface, the application calls either the routine CommPutChar() to send a single byte of data or the function CommPutBuff() to transmit a buffer of data. In either case, if the function returns a nonzero status value, an error has occurred (see the file StatusDef.h for a complete list of status code values).

Calling sequence:

```
Uint16 CommPutChar(Uint8 ui8Data);

Uint16 CommPutBuff(Uint8 *pui8Data, Uint16 ui16Length);
```

Example: Transmit the value "d".

```
Uint8 ui8Data;          // Buffer to hold data for Tx queue

ui8Data = 'd';

CommPutChar(ui8Data);
```

Example: Transmit the string "dsPIC" (without the NUL terminator).

```
CommPutChar("dsPIC", 5);
```

Index